THE BIG
BANG
THEORY

D0324873

What It Is, Where It Came From, and Why It Works

KAREN C. FOX

JOHN WILEY & SONS, INC.

Published by John Wiley & Sons, Inc., New York
Published simultaneously in Canada

ISBN 0-471-39452-1

Printed in the United States of America

10 9 8 7 6 5 4 3 2 1

For Dad and Amber, Mom, and Bill.

Thank you.

Contents

Acknowledgments ix

Introduction 1

PART I
How We Came to Believe the Big Bang Theory 7

1 The First Cosmologies 9
2 The Birth of the Big Bang 37
3 The Search for Proof 55

Interlude: Popular Reactions 81

PART II
How Good a Theory Is It? 91

4 Scientific Reactions 93
5 Glitches 119
6 Current and Future Research 137
7 The Edge of the Unknown 159

A Big Bang Timeline 175

Notes 181

Glossary 185

Bibliography 193

Index 197

Acknowledgments

A book like this depends on the work others have dedicated their careers to producing, others more specialized than I. Various cosmologists gave of their time to discuss modern cosmology, most notably Paul Steinhardt and David Spergel at Princeton and Jim Bardeen at the University of Washington. Information from discussions with Tom Siegfried, Charles Seife, and Phil Schewe—all great science writers—is also scattered throughout the book. Much of chapter 3 relies heavily on the ideas and research of Helge Kragh, professor of history of science at Aarhus University in Denmark. The standard story of how the big bang theory came to be accepted has been told so many times that it has turned into a modern myth. Professor Kragh, and historians like him, provide constant reminders that science doesn't move in the perfect progression we've all been taught. He graciously offered advice and information, and his book *Cosmology and Controversy* was invaluable. Other fantastic books that were particularly helpful: *The Big Bang* by Joseph Silk, *Blind Watchers of the Sky* by Rocky Kolb, *Afterglow of Creation* by Marcus Chown, *The Dancing Universe* by Marcelo Gleiser, and *The Runaway Universe* by Donald Goldsmith.

Any opinions on the robustness of the big bang theory itself are, of course, mine—not those of the people mentioned here.

Much thanks to my editor, Jeff Golick. The idea for this book was his in the first place, and his suggestions on how to organize the vast amount of information have been invaluable.

Special thanks must go to the Amherst College physics department. I walked into my adviser's office in the fall of 1988 and announced that in addition to majoring in physics, I was also going to be an English major. They didn't balk; they encouraged me every step of the way, and have always been supportive of my career in science writing. Thanks also to the students in the class itself. There are a few specific people—David Hall, Claire McLean, Sean Prigge, Kent Johnson, Jennifer Eden, Crista Wilson—whom I've never had a chance to thank for their four years of help. (It's not like I exactly *copied* your homework or anything. Not exactly.) Thank you to Professor Jagu at Amherst, who read and commented on an early draft of the manuscript.

My parents, as always, have been wonderful, offering time, assistance, havens in which to write, and a deaf ear to my (occasional!) crankiness while working. Patrick Mattingly deserves to be awarded sainthood for his unwavering support, ability to listen to my complaints, and timely purchasing of Hansen's D-stress whenever needed. Thanks also to many friends—Catherine, Holly, Rebecka, Misha, Noah, Lisa, Mike, Jane, Carolyn, Carrie, David, Melanie—for accepting the on-again, off-again nature of my attention span while writing this.

Introduction

In early 1998, two groups of astrophysicists made a startling announcement. The universe, which had always been thought to be expanding, and gradually slowing down, might actually be speeding up. The scientists had been watching supernovas—the explosive death throes of a dying star—at distances greater than had ever been seen before. The speeds and distances of these brilliant fireballs suggested that all of space was swelling up ever faster, accelerating over time.

This information took everything scientists thought they knew about the way the universe was supposed to work and dumped it on its head. Suddenly gravity, the force that was supposed to be holding everything in the universe together and keeping it at a manageable size, wasn't in control. Instead some mysterious energy was actually resisting gravity and counteracting it. The findings, if true, insist that physicists reexamine how they interpret everything from Einstein's relativity theory to how the universe has evolved since the big bang.

And yet, the scientific community took the new results in stride. New data had been found that contradicted old ideas, and so the time had come to adjust those ideas. As they examined the surprising news of this runaway universe, cosmologists realized that by changing some of their assumptions, this new information might actually make the big bang theory better. By incorporating this mysterious antigravity, they were able to solve another problem—that there wasn't enough matter in the universe—that had long dogged them.

This is a fine example of how a cosmological theory gets molded. Scientists are fairly confident in the outlines of their theories, but they also must remember that these theories aren't etched in stone. A theory is worthwhile only if it matches what is actually seen. If contradictory data show up, then one has to adjust the theory. And yet it's hard to resist getting overly committed to a specific idea, to keep an open mind, and to not get caught up in the dogma. This book is about that all-important open-mindedness. It is about the attempt to fashion a solid theory, the very best theory ever devised for how the universe began, and yet to be flexible enough to modify the theory when necessary.

For most of history, cosmology has been intricately entwined with religion and philosophy. How was the universe created? Why was it created? How does it work? How unique is earth, our solar system, mankind? With questions this big, one almost has to rely on answers from all three disciplines—religion, philosophy, and science—each of which uses a fantastically different method to find an explanation. While this book touches on religion and philosophy, it is primarily concerned with the scientific description, and it's important to understand the distinction.

Religious explanations, for example, do not attempt to be objective, independent from the untestable beliefs of the people. A religion's interpretations invariably use the visible culture and landscape of its society to explain the invisible forces in the world. (Babylonian myths had the universe created out of silt, much the way their homeland was made out of silt from the ocean; Native Americans told stories of the world being created by the coyotes and foxes like those living in the woods around them.) Religious stories about the origins of the universe are valuable in the ways they offer comfort and order in what can be a chaotic world. As literal truths, however, they lose their potency, since they fail to paint a picture that corresponds to what we observe.

Philosophy, on the other hand, arising in Greece in about 500 B.C.E., does offer explanations believed to be "true." It doesn't depend on any divine or magical hand to describe the world and instead offers rational explanations. But philosophy doesn't involve experiments or rigorous testing of any theories. In fact, the modern interpretation of philosophy is that it is reserved for those things that can never be tested. In the past, however, philosophers did attempt to explain the physical world around them, though, of course, no experiments were done. In fact, some philosophers, such as Plato, rejected out of hand any observations that contradicted the beautiful theories generated in the mind. Observations could deceive; math was perfect.

Science changed all that. As new theories were created about the universe, scholars began to observe and experiment and test to see how well the theories corresponded to reality. Scientific theories do not coddle our initial interpretations of the world around us. Indeed, many scientific facts are quite startling: the earth is not the center of the universe but a run-of-the-mill planet around a run-of-the-mill star on the outskirts of a run-of-the-mill galaxy; this book is made up of tiny particles called atoms and is 99.9 percent empty space; and millions of subatomic particles such as neutrinos are passing harmlessly through your body this second.

But while some facts seem counterintuitive, their power lies in that they consistently coincide with what modern tools actually observe. The facts are predictive. If we postulate a solar system where all the planets move around the sun, then when we chart Venus's movement across the sky, it must always follow the path we expect. Science today claims that Venus travels in an ellipse around the sun, and we can map how that would appear to a person standing on earth. So far Venus has always gone the right way around. If it ever disappoints, we're going to have to adjust our theories.

Aristotle, a Greek philosopher of such importance that in the Middle Ages he was referred to as simply the Philosopher, devised a theory of the solar system with the earth in the middle. The sun, the moon, and all the planets moved around the earth. To make this model correspond with what was actually observed in the sky, the planets were assigned complicated orbits of circles upon circles, which corresponded quite well to the observed paths. The system was complicated—King Alfonso X of Spain remarked in the thirteenth century that had the Lord asked him for advice, he would have come up with something simpler—but it worked. If NASA's scientists wanted to map the path for a rocket to land on Mars using ancient geocentric charts, they'd come pretty close. They might not get a mark as precisely as they currently do, but they'd probably succeed. The modern vision of the universe with the sun in the center, a combined effort of Copernicus, Kepler, and Galileo, was accepted because additional information—information that made it clear the Aristotelian version couldn't be true—came to view when new tools such as the telescope evolved.

A scientific theory always risks being overturned. While religious stories are believed out of faith, and philosophy out of the most pleasing logic, scientific theories aren't meant to be ever accepted completely. They are never proven beyond a shadow of a doubt. Just as believing dogmatically in one version of any myth or religion leads only to narrow understandings of the world, so dogmatically believing in any scientific theory is problematic.

And yet, science does attempt to seek *literal* truths, so scientists must walk a fine line, knowing their theories have value only if they're accurate, and yet knowing they are not necessarily dead on. It's much like doing a crossword puzzle. There *are* right answers, and you must trust occasional guesses if you are to get far enough to solve the whole thing. And yet you must always be willing to see one of your words as wrong, and to scrap a whole square of the puzzle you thought you had down correctly. Easy enough with a crossword puzzle, but change the issue to your fundamental view of how the world works and many people have a difficult time keeping the right mind-set.

Throughout the millennia, many brilliant people have created completely inaccurate portraits of the universe because of unexamined beliefs taken as dogma. We shall take a tour of their travails and then take a long, hard look at whether the theory of the big bang is one destined to survive or fail.

As we question the robustness of the big bang, it's important to remember that one can question a scientific theory without insisting on overthrowing all of science or the scientific method. It's a shame that this distinction should even have to be made. The worlds of science and religion are so distinct, operating under such different rules, that neither benefits from being "confirmed" by the other. They should be able to exist side by side without conflict; apparent contradictions can be dismissed if you realize they are two fundamentally different ways of relating to our surroundings.

This book therefore sets out to do two things: first, to establish the big bang's legitimacy as a scientific theory, and explain why it's been accepted; then, in part II, to explain just what problems—from a scientific perspective—are still there. We will describe the holes that still need to be patched up, walk through the experiments currently being done to hammer out the last details, and address the likelihood of whether the whole theory will ever be scrapped.

The big bang theory was suggested, partially tested, and accepted in the first two-thirds of the twentieth century. In an effort to truly prove the big bang, scores of new technologies—such as the Hubble Space Telescope—have been developed. We are in the midst of an experimental boom. With the first rounds of data just starting to come in, cosmology is beginning to turn into a testable science.

We will soon know if the big bang is just another theory in the history of human ideas.

PART I

How We Came to Believe the Big Bang Theory

1

The First Cosmologies

In the beginning, there was nothing. Well, not quite nothing—more of a Nothing with Potential. A nothingness in which packets of energy fleeted in and out of existence, popping into oblivion as quickly as they appeared. One of these fluctuations had just enough energy to take off. It inflated wildly out of control—one moment infinitesimally small, moments later light-years across. All of space and time was created in that instant, and as that energy slowed, it cooled and froze into matter—protons and neutrons and photons. This baby universe kept expanding, over billions of years, and those particles coalesced into stars and planets and eventually humans.

And that's how the universe came to be.

Or at least that's the modern version. Descriptions of how the cosmos was born, from the dramatic to the lyrical, have proliferated throughout human history. Take the Enuma Elish, the creation myth recited in about 4000 B.C.E. on the fourth day of each new year by the Babylonians as they lay prostrate before a statue of their great god Marduk.

The epic begins:

When heaven was not named,
and the earth beneath did not bear a name
nor the primeval Apsu,
who begat them,
nor Tiamet, the mother of them.

Their waters, sweet and bitter,
mingled together.
And no field was formed,
no marsh was to be seen.
When none of the gods
had been called into being,
And none bore a name,
and no destinies were ordained.
Then in the midst of the waters,
gods were created.

Lahmu and Lahamu,
were called into being.

The non-Babylonians among us may need help understanding who these divine beings are. Apsu is the sweet river water; Tiamat, the salty ocean. They come together—just the way the Babylonians would have watched a river delta hit the sea—and they create the first gods: Lahmu and Lahamu represent silt and muddy slime, the earth itself.

A child hearing these two stories for the first time would have no way of choosing which one was correct. As stories, one simply chooses the prettier version—and most would probably want to go with the poetry. But there is a dramatic difference between the two: the big bang description is more than just a fable. Science prides itself on providing physically accurate descriptions of the observable universe; it seeks literal truths about what *really* happened. The methods of religion and science are vastly different.

So how does science decide on just one theory, the accurate one, the *true* one?

The way many people like to imagine the process is that everyone keeps an admirably open mind and changes his or her beliefs when enough contradictory proof collapses an old theory. Science is cumulative, they think, refining theories over time, always getting closer and closer to the "truth."

Thomas Kuhn explained the process a very different way. A historian of science who taught at Harvard, Kuhn advanced the notion of the "paradigm shift." He said that scientists worked within the confines of their theories long after there was enough factual proof to disprove them. A new theory was embraced only when someone finally overturned the whole shebang completely, as thoroughly as upsetting a dinner table of dishes and silverware to the floor. A new theory would arise in place of the old one, with a whole new language and set of assumptions to go with it. We believed new theories not because of the preponderance of evidence, but because they "made sense" to us, much the way a creation myth might.

And then there is a third view, bluntly expressed by Max Planck, who simply said that scientific theories don't change because scientists change their minds; they change because old scientists die.

Like most philosophical truths, the answer is probably some combination of all of the above. The story of how the big bang theory became accepted certainly contains a bit of each. There are geniuses who scoffed at all previous theories and devised brand-new, innovative solutions; and there are those who refused to reject their preconceived notions no matter how overwhelming the evidence. There are people who accepted cosmology theories based on the experiments only; and there are those who simply believed them with the same faith they might bring to a religion. There are those today who say they believe the big bang model because all the experiments support it; and there are those who say there are enough holes that we shouldn't believe it just yet.

The story of how the big bang theory was accepted incorporates all these scientific styles, and it begins in ancient Greece.

EARLY BELIEFS

Imagine looking at the sky for the first time—as if you knew nothing about its workings. Every day you'd see the sun rise and then set as the moon rose up to replace it. By and large you'd think they were similar bodies, both traveling around the globe in exactly the same way.

At first, all those other bright stars in the night sky would seem to be entirely different. They could be as small as fireflies hovering just a mile or so up in the air for all you knew. The fact that the moon passes in front of them in her nightly journey will tell you they're at least farther away than the moon, but that's about all. In time you'd see the night sky move around the earth, too, all the stars moving in unison, and assume it was some two-dimensional backdrop twirling around a fixed earth at the center.

Eventually you'd notice that six of those stars didn't move in sync with the others. Each has its own erratic path. Each moves across the sky in one direction until suddenly it slows down, stops completely, and begins moving in the opposite direction for a while before finally resuming its course. Sometimes these six would even get larger, as if they were coming closer. You'd know these were different somehow, and your first guess might be that they were like the sun, just much, much farther away, circling the earth as the sun does.

And that would be about as far as you'd get. You could record daily movements of these wandering planets (the very word "planet" means "wanderer" in Greek), but your predictions about why they moved and how they were created would be based only on what you saw with the naked eye—and would be highly inaccurate.

Sure, you might realize that the sun moving around the earth would make the sky look just the same as if the earth twirled on its own axis, but relying on nothing more than your unaided vision, you really couldn't prove anything one way or the other. And the idea of the world twirling wouldn't make too much sense, because wouldn't you expect to feel the wind rushing past?

It's into this world, with just this much knowledge and this much observation, that the Greek philosophers arrived. They could map the night sky, they could predict eclipses, they could geometer their way into beautiful drawings of planetary motions, and they used all of this to devise theories about the universe, but they didn't have the tools to prove any of it.

The Greek philosophers didn't rely on experiments the way modern scientists do, but they took a giant step in that direction by rejecting religious explanations and denying that gods caused daily phenomena. Lightning bolts weren't hurled to the ground by the angry arm of Zeus, but born of natural processes, processes that could be rationally explained. Lucretius, for example, writing in the first century B.C.E., said: "Nature is free and uncontrolled by proud masters and runs the universe by herself without the aid of gods. For who—by the sacred hearts of the gods who pass their unruffled lives, their placid aeon, in calm and peace!—who can rule the sum total of the measureless?"[1]

Notice how just as Lucretius denies divine intervention he invokes the existence of gods. He's our very first example of someone who embraced a new theory yet couldn't quite let go of what he'd been taught all his life. His whole philosophy was based on the idea that nature—made of little atoms—ran independently of the divine. But jettisoning the gods altogether? That would be going too far.

The Greek philosophers came up with numerous versions of how the universe worked. (A personal favorite: Anaximander's proposal that we live inside a huge sphere with fire along the outside rim—the sun is nothing but a hole in the sphere through which we can see the fire.) But one model truly captured the imagination of the Western world well into the 1600s. This was the version described by Aristotle; his would be the unquestioned assumptions that philosophers and scientists embraced for centuries. And Aristotle got the assumptions from his teacher Plato.

Aristotle and Plato

Plato lived in tumultuous times. Born in 427 B.C.E., he grew up during the Peloponnesian War and learned his philosophy in the marketplace where Socrates preached to the young men, telling them to question the morals they'd been taught. The community elders—who'd done all that moral teaching—weren't pleased, and Socrates was soon tried and sentenced to death.

Perhaps it was his friend's death and the chaos of a war-torn nation that led Plato to seek solid truths to put order and calm into his world. Plato lived in the world of ideals. The physical world, he claimed, was merely a facsimile of the perfection created in some divine mind. Plato referred to this divinity as the *Demiurge.* We, too, could experience true reality only in our own minds: a circle can only be perfect in our thoughts, after all—draw one and it's invariably slightly off-kilter.

Plato's universe depended on these "perfections." He picked five ideal shapes and claimed that the "elements" matched them: fire was a tetrahedron (a three-sided pyramid), the sharp sides of which could cut the connection between other elements; earth was a solid cube; air, an octahedron (a solid with eight sides that look like pentagons); and water was a slippery twenty-sided icosahedron (each side is an equilateral triangle). The planets, he believed, must travel in circles with uniform motion.

No matter that there was no proof of any of this, or even that the planets zigzagged across the sky in a way that looked distinctly noncircular. Because it was a beautiful theory, Plato believed it must be true. Those who thought they could understand the stars by, gasp, *observation* were as empty-headed as birds, as he wrote in his *Timaeus:* "But the race of birds was created out of innocent light-minded men, who, although their minds were directed toward heaven, imagined, in their simplicity, that the clearest demonstration of the things above was to be obtained by sight."

Plato's blindness did not, therefore, come from an inability to see but from his disdain for what he saw. He chose to ignore what he observed. He created a theory of the universe with everything moving in these perfect circles—even though it certainly doesn't look like this from earth as the planets wander backward and forward through the night sky—and then asked others to produce a mathematical model that would fit it.*

Yes, those planets obviously traveled back and forth through the sky, so how do you create a universe run only by perfect solids and circles? Eudoxus, born in Sparta in 408 B.C.E., and one of Plato's pupils, rose to the challenge. To "save the appearances," Eudoxus described a set of planets sitting on a series of moving spheres with the earth at their center. The sun, for example, had three spheres: one to move around the earth daily; another, slower sphere, which on an annual basis moved to account for the way the sun appears to move higher in the sky throughout the seasons; and a third sphere to explain some incorrect observations of the

*At least this is so according to Simplicius, who commented on Plato in the sixth century C.E. And in fact he is commenting on a commentary on a commentary of Eudoxus's report about Plato's teachings, so as with much of what we know about the ancient Greek philosophers, we get our information at the tail end of one big game of "telephone."

time that had the sun changing position on the horizon from equinox to equinox. Eudoxus described each planet this way, adding on spheres moving at different speeds and in different directions, until there were twenty-seven spheres in total.

The model was totally wrong, of course, but the concept itself was mind-boggling. For the first time, someone hammered out a mathematical model correlated to what he saw. Eudoxus didn't think it was a description of what was actually happening in the heavens, and his model didn't perfectly fit the data, but you could use it to predict where a planet should be with reasonable accuracy. Math corresponded to reality.

Math corresponding to reality is seductive. It makes you believe the model is correct. In fact, using a theory to successfully predict an outcome in the physical world is exactly the kind of thing that gives modern-day scientists confidence in their models. If a theory doesn't fit reality, that's easy—you discount it immediately. The theory is wrong, and it's time to move on. When a theory fits the data, however, one doesn't quite know what to think. Sure, there's a chance it's right, but what about Eudoxus's rings, a theory we now know to be hogwash? Scientists must always remember how false theories have been emphatically believed in the past. This one, with the help of Aristotle, would be believed for centuries as insistently as we believe today that the earth goes around the sun—all because it by and large fit the data.

Aristotle added a twist to Eudoxus's model: he turned those ephemeral spheres into something solid. If the math worked, why couldn't it be a valid physical description? Aristotle studied under Plato for some twenty years before founding his own academy, but he wasn't as afraid of observation as his mentor was. Perhaps Aristotle's world simply seemed more stable. Aristotle was born in northern Greece in Macedon in 384 B.C.E. His father served as personal physician to the ruler, Amyntas II. Aristotle's life in Athens was largely good, as he was originally tapped to be Plato's successor.

Aristotle eventually stepped far enough away from the master that he wasn't chosen to lead the academy, but Plato's influence over his philosophy would be the rut that kept Aristotle's theories from being accurate. Aristotle inherited Plato's incorrect assumptions: Aristotle trusts so implicitly in the obvious perfection of a sphere that he never bothers to offer detailed proofs, as he does for other ideas, that the planets move in circles. In "On the Heavens" he writes: "The shape of the heaven must be spherical. That is most suitable to its substance, and is the primary shape in nature." So there.

The spheres in Aristotle's cosmology stemmed from Eudoxus's, but they needed some jury-rigging to become a physical reality. Aristotle devised a system whereby each sphere forced the spheres inside to rotate with it. Consequently he had to add spheres not only to account for the oddities of each planet's rotation but also to negate movement from the sphere above it. In the end his version had fifty-six spheres and roughly accounted for most celestial movements. It failed to explain a couple of crucial points, however, including why the planets periodically shone brighter and larger, as if they had swung closer.

But these shortcomings didn't derail it. Aristotle's description of the universe caught the imagination of mankind in a way that no previous system did. Christians in the Middle Ages would take it as a perfect model of a God-governed cosmos, since Aristotle's universe included an outermost sphere, which they thought was "heaven."

The way it works is this: Everything on earth is made of the four elements, earth, air, fire, and water. All these elements seek to be at rest in their "natural" state. Earth and water seek rest by moving in straight lines "down" (toward the center of the earth), while fire and air naturally move in straight lines "up." A rock that moves horizontally—say, by being thrown through the air—does this only because an outer body has forced "unnatural motion" upon it.

At the beginning of time all the earth and water naturally fell down toward the center of the universe, clumped together, and formed the spherical globe on which we live. This automatically proved there were no other universes out there, because that would imply two conflicting "downs"—an absurdity to Aristotle. (Today we know there are billions of "downs" in the universe. "Down" is always in the direction of the greatest gravity nearby. For us it's earth, but it could be the moon or the sun or a black hole.)

Once you made it out of our atmosphere, however, everything changed. Starting with the moon, the universe was made of a heavenly material known as "ether." While everything on earth moves in straight lines, ether naturally moves in circles—divine things simply must move in the "perfect" shape, after all. There were spheres for the moon, the sun, the planets on out to the stars, and then beyond that was an Unmoved Mover—the divine force that moved the outer sphere that set each of the next fifty-five moving. Interestingly, Aristotle created this complicated mathematical model of moving spheres, and yet at times he writes that each planet is moved by an individual divinity (medieval Christians would later update these creatures to angels). While Aristotle created a cosmology that doesn't *require* divine intervention, he nevertheless insists upon it. He couldn't escape an ingrained belief that even a rational universe was a place filled with gods and hallowed beings far more powerful than mankind. This was a lovely model for the early Christian world to hold on to, and religion and cosmology would remain intricately entwined for the next thirteen centuries.

While Aristotle's cosmology became dogma, one other ancient gets credit for giving the theory the grounding needed to make it so stable. Aristotle's spheres still didn't correspond well to everything observed; hammering out a mechanical model took a true mathematician: Ptolemy.

Klaudios Ptolemaios lived in the second century C.E. and created the most comprehensive model of the planets to date. Ptolemy knew that as the planets moved around the earth they appeared to stop and go backward for a while. (We now refer to this apparent jog in a planet's travels as a "retrograde"—the planet does not, in fact, ever move backward, it simply looks that way since we watch the planets moving around the sun as we ourselves move around the sun.) To account for this, Ptolemy added a twist to the orbits. He claimed that each planet moved in tight little circles, and it was the very center of *this* circle that moved in a gigantic circular orbit around the earth. This little extra orbit was named an *epicycle.*

Ptolemy's model corresponded quite well with what we see. After all, if you're trying to figure out where the sun is going to be at any given moment, you can do that fairly well whether you assume the sun stays still and the earth spins or that the sun moves around the earth. But an inaccurate model is bound to have some anomalies. The biggest weakness in Ptolemy's model was his complicated description of Mars. Mars, orbiting the sun so close to us, has the hardest path to pin down if you're afflicted with an incorrect theory. Mars just refuses to move in perfect circles. Mars was destined to be the first nail in the coffin for Ptolemy and Aristotle.

INTRODUCING OBSERVATION

Tycho Brahe had a love-hate relationship with his brash young assistant Johannes Kepler. Over the year and a half they worked together, Kepler and Brahe fought, Kepler left in angry fits or Brahe would order him to leave, then one or the other would beg to work together again. For some reason they needed each other. Brahe's observations of the universe were unsurpassed throughout the world. Kepler wanted access to them to devise better descriptions of how the heavens worked. Brahe, in turn, needed Kepler's brilliant math skills to help prove his own theories.

But they rubbed each other the wrong way, and Brahe wanted to keep him occupied and out of his hair. "Describe for me the orbit of Mars," he told Kepler in 1600, knowing full well that all the greats of history had wrestled with the prickly warrior planet and come up short.

"Give me a week," replied Kepler.

Kepler was overconfident. It would take him eight years to hammer out the orbit, but when he did, he turned Ptolemy's model on its head. He described the way the planets move with three laws that are still taught in introductory physics classes around the world.

Brahe and Kepler's names aren't well known outside the world of astronomy, but their contributions to science were as crucial as those of the more famous Copernicus and Galileo. In fact, some historians argue that the vision of a sun-centered cosmos that Copernicus devised in the 1500s wasn't so revolutionary—although Copernicus championed a moving earth, he clung unquestioningly to Aristotelian ideals such as circular motion. He wrote: "It is altogether absurd that a heavenly body should not always move with a uniform velocity in a perfect circle." In fact, Copernicus tried to connect to the ancients by writing his most important treatise, *De Revolutionibus Orbium* (The Revolutions of the Celestial Orbs), which echoed Ptolemy's *Amalgaest,* with each chapter of his book correlating to a chapter in Ptolemy. Moreover, Copernicus cited pre-Aristotelian philosophers such as the Pythagoreans to support his ideas for a sun-centered universe. To overturn Aristotle, he reached back even farther in time—not what one would call really taking a leap into the unknown.

Not to belittle Copernicus—he certainly provided a jump from the status quo. He also used a new concept to choose his theory: simplicity. Or even, one could say, beauty. The austerity of the simple Copernican system gave it an aesthetically appealing quality missing from Ptolemy's rings within rings. In modern times, scientists have almost deified this simplicity concept, known as Occam's Razor. If two theories fit the data, choose the simpler of the two. If you can explain the cosmos with just a few orbits instead of all those epicycles, then stick to the former.

But, alas for Copernicus, his simple ring system of planets orbiting a slightly off-center sun didn't correspond to reality substantially better than Ptolemy's model. Sticky Mars was still out of whack. Copernicus's *De Revolutionibus Orbium* was edited by a man named Rheticus, who, legend has it, became so frustrated with mapping the path of the red planet that he called upon the spirit world to help. A demon appeared, threw him against a couple of walls, and shouted, "*Thus* are the motions of Mars!"

Until someone decided to map the orbit of Mars first and *then* determine the math that described it instead of the other way around, no one was going to produce a complete model of the sky. That was not something Copernicus, firmly enmeshed in the philosophies of his day, was capable of doing. It was Brahe and Kepler, with their willingness to really observe what they were studying, to insist that the observations match their theories, who nudged cosmology a little closer to modern-day "science."

That these two characters—for they were definitely both characters—had the chance to come together and collaborate is almost beyond belief. Brahe was of Danish nobility, Kepler of the German lower class, but together they provided the most accurate depictions of the stars until that time.

Brahe was a nobleman born in 1546, who by family connections was in favor with the Danish king, Frederick, who gave him an entire island in the Danish Sound on which to set up his astronomy lab. Brahe's microcosm, which he named Uraniborg, consisted of a chemical lab, a printing press, a paper mill, a prison, a game preserve, a library, a castle, and, of course, the largest observatory ever seen. Brahe built it all with money from the Danish coffers (30 percent of that year's Danish budget, to be exact—who says NASA isn't a bargain?), and he then lived off the income generated by his tenants and the various factories. Living in high style, Brahe even had a jester dwarf who sat under his chair at dinner and begged for scraps.

But Brahe's love for the stars was genuine. He knew them like the back of his hand—they were as familiar to him as the landscape around our homes is to most of us. So the sudden appearance one day of a bright star where none had been before threw him off the way the sudden appearance of a fifty-foot oak tree in our backyards would affect the rest of us. Ever since Aristotle, everyone had agreed that the heavens did not change. They were immutable, perfect, the way divine things should be. So convinced was everyone of this that when Western astronomers had noticed various comets throughout the centuries, they just assumed the comets were below the moon, the only possible place such change was allowed.

And yet, one night—it was November 11, 1572—Brahe looked up into the sky after dinner and there was a new star, smack in the middle of the constellation Cassiopeia, brighter than all the rest. For the next few weeks it was even visible during the day, slowly fading over the course of the next year and a half until it disappeared.

Brahe had witnessed a supernova—the gigantic explosion that occurs during a star's last death throes. Of course,

Brahe didn't know that at the time. All he knew was that he had witnessed the impossible: the perfect, ever-constant, starry sky had just changed. He may not have realized the implications immediately, but by the time he spotted a comet in 1577, Brahe knew what to look for. Making the most precise measurements anyone had ever seen, and presenting them—in an unprecedented move—with all his data and all the possible errors, Brahe showed that the comet was far above the moon. Not just change in the heavens, then, but a body slicing right through Aristotle's crystalline celestial spheres.

With the level of detail presented, Brahe's destruction of the Aristotelian universe was hard to dispute. The celestial spheres had been smashed, never to return. That was one of Brahe's major contributions to cosmology, but his largest may well have been simply his incredible attention to detail. His data were orders of magnitude better than anyone else's. (Remarkably, Brahe never once looked through a telescope—an instrument that arrived in Europe several years after he died.)

The minutiae matter in science. It is the minutiae that disprove a false theory and support a true one. And that is a lesson learned from Tycho Brahe.

But Brahe, like so many before him, was also a slave to his preconceptions. Unable to convince himself that the earth moved, and unable to make his observations match a geocentric universe, Brahe drew a solar system in which the sun and the moon orbited the earth, while all the other planets revolved around the sun.

The glory of Uraniborg didn't last long. In 1597, after a new king had claimed the Danish throne, Brahe ceased to be a royal favorite. He picked up his instruments and left. By 1599 he had taken on a new job: imperial mathemetician to Holy Roman Emperor Rudolph II in modern-day Czechoslovakia. It was here that he met Johannes Kepler.

Whereas Brahe was near-royalty, Kepler grew up in the poor neighborhoods of Weil der Stadt, a town in Germany. Born in 1575, Kepler was cranky, sickly, and dirty (he took exactly one bath in his entire life and claimed it made him ill). He nearly died of smallpox when he was four, and the litany of illnesses continues from there. He was nearsighted, and had multiple vision his whole life. He didn't make many friends and got into fights with his classmates. His father was a mercenary soldier, fighting for the best pay; his mother had nearly been burned as a witch in their hometown, so now she stayed with her husband in the battle camps. Kepler was raised by his grandmother, whom he described in his diary as "restless, clever, and lying, of a fiery nature, an inveterate troublemaker, violent." Kepler always kept meticulous diaries. They are a combination of harsh portraits of the people around him and detailed descriptions of his life and contributions to science. (This makes Kepler one of the few scientists in history for whom we have a description of how he lost his virginity. He was twenty-one and—surprise—it made him sick.)

And from these inauspicious beginnings sprang genius. Young Johannes Kepler was lucky enough to be born at a time when the new Lutheran religion sweeping Europe gave rise to numerous public schools. Kepler received a free education, and he eventually attended a Lutheran university in Tübingen. It was there that he first discovered Copernicus, learning of a sun-centered world from his mentor Michael Mastlin. So it was that Kepler was one of the first astronomers thoroughly indoctrinated in a heliocentric universe—he never doubted that the sun stood still while the earth rotated around it.

After college, Kepler took a job as a math teacher, in which position his ability to win friends and influence people continued at the same abysmal rate. (First day of classes, his second year: Kepler walked into an empty classroom. Not a soul had registered for his class.) While he didn't do much for his students, that classroom did a great deal for Kepler. In the middle of a lecture on geometry, Kepler had what he would later describe as one of the most profound thoughts of his life. It's a thought we now know to have no scientific merit, but to Kepler it was his greatest achievement. And while it isn't valued today, it was the first step that led him to three beautiful equations that describe the whole solar system.

Kepler had long wrestled with the question of *why* there were six planets—not seven, not five.* In a flash—as he stood in front of that class—Kepler thought that the planets themselves might be organized according to the five Platonic "ideal" shapes. First came Mercury's circular orbit. Next, imagine an octahedron (with eight sides) positioned around that circle the way any regular shape can be snugly fit around a sphere. Put another sphere directly around that, and you get the orbit for Venus. Next an icosahedron (twenty sides), the sphere of earth's orbit; then a dodecahedron (twelve sides), Mars; then a pyramid, Jupiter; then a cube; and finally Saturn. This pretty arrangement of geometry now showed *why* each planet's orbit was just so far out and no farther.

*Today, of course, we know there are nine planets (if you include Pluto, which isn't exactly a planet like the others), but in those days six it was.

As it happened, the ratios among the planets didn't perfectly match up to the ratios of the solids, and these days we know that there are more than six planets. We also know that invoking the Platonic solids as a logical "reason" for why there were six doesn't make much sense to begin with. But by and large Kepler's vision of the solar system linked up to what he observed—and that alone was pretty novel for a medieval astronomer.

Perhaps sensing a kindred spirit who understood the value of matching observation to theory, Tycho Brahe invited Kepler to work with him in 1600. Upon studying Brahe's details of the Martian orbit, Kepler realized that the planet did not move at a constant speed. (Aristotle's spheres and unchanging heavens were already discredited; now his theories of uniform motion were under attack.) But that revelation alone wasn't enough to account for Brahe's precise data. It was a struggle that took years, but finally Kepler had to face the facts: Aristotle's perfect circles must go. Kepler showed that the planets moved in egg-shaped ovals, in ellipses.

Not that anyone listened. One who dismissed it outright was Galileo Galilei. He was the only other contemporary scientist who believed that the earth revolved around the sun, but he basically ignored all of Kepler's ideas.

THE FIRST SCIENTIST

Galileo is often credited with being the first scientist, the first experimenter. And it's a valid title—his work with pendulums and gravity and momentum all revolved around experiments. But when it came to the sky, he had his failings. He refused to admit that the close correlation between Kepler's ellipses and observation discounted the theory of circles. Possibly this was mere pride, a way to discredit the only other champion of a sun-centered universe so he could hoard all the glory. After all, if there's one thing we know for sure about Galileo it's that he was extremely sure of himself, even when he was spectacularly wrong. He once wrote, "It was granted to me alone to discover all the new phenomena in the sky and nothing to anybody else. This is the truth which neither malice nor envy can suppress."

While he denied Kepler, Galileo wasn't normally so dismissive of observation. It was observation that made him a legend. At the beginning of the seventeenth century, someone invented the telescope in Holland. No one can quite agree on who it was, but it was Galileo who made the telescope great. Fitting two lenses together on either end of a long tube, Galileo pointed that telescope to the sky and saw farther than anyone had seen before. (He also sold the patent rights to the Venetian government—patent rights to an instrument he didn't invent, but no matter . . . he got himself a nice raise in the bargain.)

While others used the telescope to track the movements of ships or approaching armies, Galileo focused his tubes upward. He saw more stars than had ever been seen. He saw planets—full circles instead of mere sparks of light. He saw rings around Saturn. And, most of all, he saw phases of the planet Venus. Just as the moon appears to cycle from new moon to full because it circles the earth, so Venus appears thinner and wider in the sky as it circles the sun. It was in deciphering this pattern that Galileo saw proof that Copernicus had been right all along.

Unfortunately, the church wasn't so quick to agree. When Copernicus lived, thought, and wrote, the church was actively involved with philosophy and science, allowing and even encouraging new ideas and detailed analyses. But nearly a century later, the pope had suffered an unprecedented attack: heeding the cry of Martin Luther, many Christians had rejected the papacy. Having seen what dissent could do, the Catholic Church became infinitely less tolerant. This sun-centered stuff—directly contradictory to what was said in the Bible—must stop. (It didn't help that in Galileo's book defending heliocentrism, a quasi-fictional account of three men discussing the universe, for which he did in fact receive permission ahead of time from the Catholic Church, he attributed the church's point of view to the stupidest, dullest man in the book, thus appearing to mock papal views.)

In a trial still famous today, Galileo was summoned before the pope's inquisitors and made to denounce his work. Opting to keep his life, Galileo signed papers stating that the earth most definitely, unequivocally did not now move, and never had. He was sentenced to house arrest for the rest of his life.

Legend has it that as he stood up and walked out of the church he muttered, *"Eppur si muove"* (Yet it moves). Whether or not Galileo said the actual words, the earth's movement was something no one could deny for much longer. The cracks in Aristotle's theory were too large. It would soon fall apart.

NEWTON

What we have so far are brilliant men—for they were all men at this point—each able to see a piece of the puzzle. As in the fable of the blind men exploring an elephant—in which one man touches the trunk and thinks that's what an elephant looks like, while the others make different decisions based on touching the leg or tail—each of these greats had an awesome understanding of his one piece of the animal. Their fault lay in denying the other's bit. Copernicus envisioned a sun-centered universe but refused to let go of Greek philosophy. Brahe was willing to ditch the crystalline spheres but wouldn't let go of a geocentric universe. Kepler abandoned Plato's circles but couldn't abandon the Platonic solids. Galileo accepted Copernicus but refused to embrace Kepler's ellipses. Each took a giant step forward but couldn't pull everything together into a whole. It was Newton who, a generation later, took a step back and saw the whole elephant.

Like the geniuses before him, Newton was an arrogant man. Perhaps the courage to capsize centuries-old theories must stem from a certain amount of hubris. Perhaps confidence in the face of criticism is the most important ingredient for toppling dogma. After all, no matter how much we try to make our choices based on logic, no matter how much mathematical proof there is for a theory, humans tend to "believe" a scientific theory the same way they believe in a creation myth—somewhere in their gut. So overthrowing a universally accepted theory takes more than data; it takes superhuman conviction.

Newton didn't have as much of an uphill battle to convince his peers as those who came before him. In his day, many already accepted the theories of Kepler and Galileo. Newton lived to see his ideas embraced, and so he spent the last thirty years of his life basking in England's admiration of their real-life hero.

Newton was born on Christmas Day in 1642. His mother saw to it that he got a first-rate education—Newton was the first person in his family who could write his own name—but other than that, she by and large neglected him. Newton was raised by his grandmother, while his mother and stepfather lived in another village down the road. Newton's estrangement from his mother seems to have affected the rest of his life—he never made close connections with the people around him, never married, never felt comfortable in society. He turned whatever energies might have gone into personal relations inward. Once caught up in a puzzle, he would work around the clock, ignoring everything around him, including the most basic of problems such as hunger and exhaustion.

The ferocity with which Newton attacked problems in science and math is mind-boggling. In a scant two years, beginning when he was twenty-two, Newton created calculus, learned that white light was a superposition of all the colors in the rainbow, and—most interestingly for us—devised the theory of gravity.

Newton began to think about gravity because he couldn't figure out why the moon merrily continued to orbit the earth instead of flinging off into space. (Think about whirling something like a yo-yo around and around your head—the only thing keeping it moving in circles is the fact that you're holding the string. The moment you let go, it will shoot off in a straight line. Contrary to Aristotle's assumptions, Newton knew that nothing *naturally* traveled in a circle.) Sitting under an apple tree one day—you know the story—Newton saw an apple plop to the ground. He realized that the same force that pulled on fruits and humans and animals might extend all the way up to the moon. This force of gravity could bind the moon to the earth as firmly as a yo-yo string keeps that yo-yo whirling around your head.

It would be some time before Newton realized that the implications went farther than the moon's path. Incorporating gravity into Kepler's equations, Newton realized he could produce mathematical descriptions of gravity that matched up perfectly with elliptical orbits. Gravity governed not only the moon but all the planets.

Perhaps it went even farther than that. In the end Newton proposed a theory of universal gravitation: Every single object in the universe, he claimed, attracts all others through the force of gravity. Small bodies, such as this book, create such weak gravity that you don't feel it. But take something as massive as a planet, and you're going to be pulled to its surface like a magnet to a refrigerator. It would be more than twenty years after that falling apple before Newton published, in 1687, a complete description of his insight, in *Philosophiae naturalis principia mathematica* (Mathematical Principles of Natural Philosophy).

Today we refer to the work as the *Principia,* and it is still considered one of the most revolutionary books of all time.

Starting from a few basic principles, incorporating some math, and throwing in completely new insights about the nature of matter, Newton explained everything that was known about movement. From how a rock moves through the air, to why friction slows down a rolling ball, to why the planets orbit the sun, Newton's laws explained it all.

Of course, universal gravitation had implications for cosmology as well. If everything is attracted to everything else, then planets and stars and the earth should be constantly in erratic motion, constantly pulled one way or another. Our universe shouldn't be stable, and yet by all appearances it was. (It never occurred to anyone that the universe *wasn't* stable, completely static. Figuring out that the universe expands would be crucial in developing the big bang theory, but the idea was considered so bizarre that even Einstein refused to believe it at first.) To solve the problem, Newton claimed that the universe must be infinite. If it was infinitely large, it was just possible that the force from every body in space perfectly balanced the force from every other body. Pulled an equal amount in all directions, each planet would stay put. Newton granted that this was the equivalent of balancing an infinite number of needles on their points, but nothing was beyond the ability of God.

Yes, God was still a large part of Newton's vision. He believed that such universal order could be explained only by the hand of a divine being. Moreover, such perfect stability would require God's hand constantly intervening, constantly tweaking, moving this star slightly to the left, this planet slightly to the right to maintain the perfect balance. Newton knew the difference between "science" and "religion." Science had to be backed up by proof and observation—you couldn't just make up random hypotheses; they had to correspond to physical reality. But it never occurred to him that religion wasn't as important a force in the universe as his neat gravity. God and science didn't clash at all, but worked in harmony.

BEYOND THE SUN

As all these scientists described the solar system, it never occurred to them to track much farther. For most, the stars were no more than a backdrop—a fixed, two-dimensional tapestry against which everything else moved. On his drawings, Copernicus merely labeled an outer sphere of stars; he never commented on what might be beyond it. A less well-known British astronomer, Thomas Digges, was the first to expand the heavens. In 1576, he translated passages from Copernicus into English and drew a famous picture showing stars beyond Saturn, extending out in all directions. It was the first suggestion in medieval times that the universe might spread in three dimensions past the planets.

With bigger and better telescopes, scientists began to map the galactic neighborhood. Having understood our home and our street, it was now time to move on to the whole block, perhaps the whole city. Of course, that's a lot like having to map a town while being forced to sit on your front porch with a pair of nice binoculars.

It's a wonder, when you think of it that way, that we have the hubris to think we really know our way around the universe at all. After all, if you see something green at the farthest edge of your visibility, that doesn't mean it's a plant. It could be a billboard, for all you know. Scientists must maintain a constant awareness of the limitations of their tools. On the other hand, you can do a fair amount of observation while sitting on your porch.

2

The Birth of the Big Bang

It's hard to pinpoint the exact moment when the seeds of the big bang idea took root. The history of cosmology is filled with stories of scientists working alone, rediscovering facts someone else had already determined ten years earlier but hadn't received much notice for. As various scientists picked and teased at parts of the puzzle, each came to important conclusions, but, working in a vacuum, each found it hard to put all the pieces together.

Certainly modern cosmology would never have gotten off the ground without the pioneering work of astronomers throughout the eighteenth and nineteenth centuries who devised new techniques for analyzing the heavens. (Stuck on their front porches, analyzing their neighborhoods, the scientists of the past three centuries devised ingenious ways of observing their world. Galileo's first telescope, for example, was only about as good as a modern-day pair of binoculars. That past astronomers did as much as they did with the tools they had is awe-inspiring.) There was William Herschel and his son John, who traveled the globe and mapped the sky more comprehensively than had ever been done before, cataloging all those fuzzy blotches on the sky—many of which would turn out to be other galaxies. There was Joseph Fraunhofer, who ran starlight through prisms and other lenses to separate out the various wavelengths of light—and was thus able to determine what types of atoms were burning at the heart of each body. (This process showed conclusively that the moon and other planets did not generate their own light but merely reflected sunlight—the spectra from each of the planets matched the sun's exactly.) There was Christian Johann Doppler, who in 1842 discovered that light, sound, or anything that moves in a wave changes frequency as it travels toward or away from the observer—a finding that would eventually be used as a technique to measure the velocities of stars as they speed away from our solar system. By the turn of the twentieth century, more scientists began to tackle larger questions, such as how our universe formed or whether it had always existed, but such scientists were few and far between.

All of these helped lay the groundwork. But modern-day cosmology really got its start in 1905 with a young man sitting at a cluttered desk in a Swiss patent office: Albert Einstein.

THE THEORY OF RELATIVITY

Albert Einstein landed at that patent office in Bern after a desperate attempt to find a job—his fiancée, Mileva Maric, was pregnant and he needed money. Einstein had met Mileva when they were in university together in Zurich. They were two of five students who enrolled the same year to become math and physics high school teachers. Contrary to the popular notion that Einstein was a lousy student, he actually did fairly well at school—though Mileva's grades were better, a trend that continued until they took their final examinations, which he passed and she didn't.

At school they fell in love and decided to marry, much to the torment of both their parents. But theirs had all the intensity of a college romance: they were convinced of their destiny together and shared a love for science. In Einstein's letters to Mileva in 1901, a time when they were living with their parents, he talks as much of "our theory" and his hopes to have "our papers published" as he does of the impending child. Mileva gave birth to a baby girl named Lieserl while in Serbia at her parents' home. As far as anyone knows, Lieserl never met her father and was never heard from again. In fact, Leiserl's existence was only discovered in 1987, when the above-mentioned letters were found. Speculations range from that she died as a baby to that she had a genetic disorder and was given up to foster care, but no one knows for sure.

Regardless, Lieserl had a hand in landing Albert Einstein in Bern, where in 1903 he finally married Mileva, a relationship that, despite all the intense love letters of those university years, would disintegrate completely by 1914. The patent office seems to have offered Einstein the perfect milieu to work on his own ideas. It was here that he developed the first version of his great theory: the Special Theory of Relativity.

At its heart lies an experience we've all had. Imagine driving along in a car at a steady clip, when the driver jams on the brakes. You naturally fall forward, and if you know anything about Newton's law of inertia ("A body at rest tends to stay at rest; a body in motion tends to stay in motion") you rationalize it by realizing that you keep moving forward simply because you've *been* moving forward. The car may have stopped immediately, but that doesn't mean you will—it takes you a moment to come to rest. This analysis assumes that motion is defined in terms of some specific thing that is at rest: the road, a tree on the side of the road, the earth itself. If you move in relation to the road, then you're moving. If you stand still along with the road, then you're at rest.

Not so, said Einstein. If you sat in that car with a blindfold on, you wouldn't know you were moving with respect to the road. You'd feel as if you were sitting still. Then, when the car suddenly stopped, you'd feel as if some force was pulling you forward. And here's the clincher: there actually *is*. Or at least that is just as valid an interpretation of what's going on as anything else. All the rules of physics must work whether you think of yourself as moving or as being at rest with a moving road speeding along underneath. Physicists say there is "no preferred reference frame." You can't say that the reference frame of thinking you're at rest when you stand on the side of the road is any more valid than the reference frame of thinking you're at rest when you're sitting in the car.

That's a simple starting point, but the repercussions are huge. As Einstein built on his theory, built it up into what is now called the General Theory of Relativity—presented to the scientific community in 1917—he realized there is no "absolute space" or "absolute time." There simply is no vantage point whereby you can get a good, objective look at anything and say for sure: "gravity pulled the ball toward the earth" as opposed to "gravity pulled the earth toward the ball." And since the laws of physics must hold true in all cases, Einstein went about changing the very nature of what "gravity" meant. Gravity wasn't just some neat force pulling you toward the earth, as if you were tethered by a string. No; gravity actually was a curve in space—massive objects such as planets curved space the same way someone standing on a trampoline warps the surface of the tramp. Anyone else who gets up on the trampoline will slip toward the depression the other person has made. A planet does the same thing: it creates a warp in space, and anything nearby slides toward it.

The theory of relativity was accepted fairly quickly, considering how ingrained Newtonian mechanics had been for centuries. It helped that most scientists already knew that Newton's gravitation laws had just a few problems. Minute details were off, such as the predictions of how the orbit of Mercury changed over time. (The fact that the change being discussed was less than one degree every eight millennia shows how vastly telescopes had improved since Galileo's.) Since a number of people had already tried their hand at modifying Newton, and since Einstein's math made so much intuitive sense—not to mention that it almost exactly matched that ever so slight oddity of Mercury—relativity was embraced fairly quickly.

A great experiment on May 29, 1919, measured the way starlight bent as it traveled through the "warped space" around the sun during an eclipse, since it was only during an eclipse that one might be able to *see* starlight in the daytime. The light was deflected exactly as Einstein predicted. This was the first experiment to support the theory of relativity, and it often gets credit for being the ultimate proof that convinced the scientific community. Certainly it was the moment of truth. The *Times* of London printed a headline: "Revolution in Science, New Theory of the Universe, Newtonian Ideas Overthrown, Momentous Pronouncement, Space Warped."

But the fact is that much of the scientific community had already gotten used to the idea and had previously embraced relativity. (And considering the expense of getting the equipment to the cocoa plantation in Africa where the team of English scientists did the experiment, you have to figure that they were fairly confident of the result.*) Einstein himself was asked what he would have done had the results of the eclipse contradicted relativity. He responded that had such a thing occurred the mistake would have been God's, not his: "The theory," he said, "is correct."

It is scientific sacrilege to suggest that one might embrace a physics theory using the same methods one uses to embrace religious faith . . . and yet the theory of relativity is an example of science that was accepted as much because it resonated within the brains and hearts of the people hearing it as because of the experimental proof.

---

*Of course, such confidence is not necessarily an experimental scientist's best friend—one would like to go into such an important experiment without assumptions about what's going to be found. There's some indication that the results did not unequivocally support relativity and yet were seen as confirmation.

Accepting the new theory of gravity easily is one thing, however; agreeing on what implications this held for the evolution of the universe is quite another. That would take another forty years.

Theories of the Universe

Einstein dutifully set about applying his new rules of gravity to the entire cosmos. By the 1920s he had a fair idea of the basics. He knew that Newton's universe, with its stars and planets kept perfectly balanced in place by gravity, required a center. Our galaxy, according to Newton, was an island in an infinite amount of space. The main complaint about this universe was that as light sped off into the void, it could never return. Thus the universe must be constantly losing energy, constantly diminishing. But Einstein created a universe that had no center, was finite, and was stable. A universe where no energy would be lost.

Remember, Einstein said that mass actually warped the space around it. All the stars and planets form a curved space, a space that keeps turning in on itself. This means that the universe might be closed around on itself: walk in one direction long enough and you're going to come back to where you started, just as when you travel around the globe. This means that the universe is finite but still doesn't have an "end"—you may travel forever in one direction and never come to a boundary. Of course, relativity predicts this curvy space, but not the exact shape. The shape of space, the kind of curves, the size, all of these depend on what's actually in the universe—mass, after all, defines how it's going to curve. No one knows exactly how the universe twists around on itself. Attempts to determine its shape continue to this day.

Specified shape or not, this unbounded, finite universe made sense to Einstein. He accepted and embraced it. But elsewhere he discovered a dramatic problem: The equations described a universe that was expanding.

Not once in history had a philosopher, a scientist, or a religion supposed that the universe was anything but static. Sure, the jury was out on whether it began at some point in time or had always been, but those who favored a "beginning" still assumed that the universe had been created exactly as it now stood. An expanding universe was so bizarre, so absurd, that Einstein denied the possibility outright.

When he presented his theories about the universe, he announced that there was still a kink or two to be worked out, and he threw in a fudge factor, something called a *cosmological constant,* that would hold the universe's expansion at bay. Some mysterious force, thought Einstein, must repel the force of expansion, keeping it contained and in a steady state of stability.

This would, Einstein said later, be his "biggest blunder." A believer in math over the vagaries of the mind when it comes to finding truth, Einstein nevertheless let his preconceived assumptions affect his interpretation.*

As it was, there were several scientists trying to create a workable, stable model of the universe from Einstein's equa-

*As discussed in later chapters, the cosmological constant, or a variation thereof, has worked its way back into current equations, so it turns out that Einstein may have been mistaken that he was mistaken.

tions. Willem de Sitter, a Dutch astronomer, tried a slight variation: a universe with absolutely nothing in it. While this was a bit off the mark, since we're fairly convinced that the universe *does* have matter in it, his model of the universe had some interesting features that would make their way into the big bang model. De Sitter's universe did expand over time, but at a constant rate. Since the movement was always the same, the basic state of the universe itself had always been the same—it did not have a beginning or even a "childhood." This universe could still be considered "stationary," since it did not fundamentally change over time. The movement nevertheless led to a prediction that light streaming to earth from the stars would appear slightly red, something that was actually seen through telescopes. It also suggested that since there was an upper limit to the speed at which anything in the universe moves (the speed of light), there were parts of the universe we could never see and could not interact with in any way. There was a "horizon" to the universe, much the same as the horizon on earth, past which we can't see despite knowing that the globe keeps going. Both reddened starlight and an event horizon would make their way into modern cosmology.

In Russia, a meteorologist named Aleksander Friedmann was also producing new models of the universe. He had taught himself relativity for the sheer joy of it, rushing to understand the new laws the moment he first heard of them. Wrapped up in what he saw as the beauty and simplicity of the math, Friedmann refused to alter the equations as Einstein had done simply to stop expansion. Friedmann didn't mimic Einstein's cosmological constant and didn't mimic de

Sitter's trick of taking out all the matter. Instead Friedmann produced a model of the universe that expanded over time at different rates. This universe began as a baby universelet at some specific point in the past and then evolved into the universe we see today.

When Friedmann published a description of his model in 1922, Einstein wrote a single paragraph to the journal in response, saying Friedmann's math was wrong—but within a few months Einstein had to retract it. Friedmann's math was impeccable, and it was Einstein who had made the mistake. While Einstein published a statement saying he'd been wrong, he still didn't like the concept. He clung to his cosmological constant. Friedmann was destined to die young, three years later, at thirty-seven, and so never lived to see his theories accepted.

While Friedmann's models were the first to truly predict a beginning to time, it was a Belgian priest who first received attention for his big bang model. Georges Lemaître was a deeply religious man who studied engineering, physics, and math before he entered the priesthood. For him, the logical consequence of Einstein's theories—that the universe had a beginning, and grew over time—jibed with his religious beliefs. He had no drive to mathematically jury-rig his way out of the problem. He had never heard of Friedmann, but he came up with a similar solution. Exploring the consequences of an expanding universe, he followed the model backward through history. He arrived at a single point in space and time when everything, every atom, every photon, every bit of energy, was coalesced into a single spot. All of matter, claimed Lemaître, was once squashed down into a "primeval atom." From this amazingly dense first particle sprang our entire universe.

When Lemaître presented his primeval atom concept publicly, Einstein was forced to agree that the math seemed to work, but he claimed that Lemaître's physics were nevertheless "abominable." Again, the mistake was Einstein's; some ten years later, when Einstein had finally come around to the expansion idea, he said that a Lemaître lecture on the subject was one of the best he'd ever heard.

All of these scientists sat with their pencils in hand, scratching out math on bits of paper. They were what we'd call theorists today, though the word wasn't much in vogue then. They took what was already known about the universe and tried to make their equations fit. They left it to others to peer into the sky and collect observations to confirm or deny the math.

More than that, these theorists themselves didn't always think of the physical reality that corresponded to their math. The concept of a beginning of the universe may have intrigued them, but it also seemed fairly abstract. To truly nail down, say, an age of the universe, seemed rather fanciful. (Lemaître was probably the only one who really believed that his description of a primeval atom was a true description of a physical history.)

So there weren't too many people rushing to their telescopes to tackle this expansion problem. No one thought to jump into the fray and actually *look.* Once again, certain assumptions in history were so strong—in this case that the universe was static—that it took someone of unique personality, someone with the confidence to buck tradition, to overturn the paradigm. That person was Edwin Hubble.

THE FIRST EVIDENCE

Edwin Powell Hubble, born in 1889, was of a distinctly American bent. He went to college at the University of Chicago, studying astronomy and math. He was athletic, confident, and charismatic—everyone was charmed by him. American science was just coming into its own, and this overconfident, gregarious personality would become something of a stereotype of the great U.S. scientists over the next few decades. Hubble attended school when it was just becoming acceptable for scientists to study in the United States instead of at the traditional German or English universities. As it was, Hubble did get in a year at Oxford in 1912 when he was a Rhodes scholar. His adviser wrote in his report that while Hubble would get an A, "I didn't care very much for his manner." Hubble's charisma must have seemed overbearing to the more reserved British professor.

It was in 1919, after a stint as a teacher and after serving in the military during World War I, that Hubble finally landed at the observatory where he would make his name. Mount Wilson, an observatory in California, hired him at $1,500 a year, and Hubble began to work, studying the spiral nebulae he'd written about in his dissertation.

These nebulae looked like fuzzy clouds through a microscope, and no one was quite sure what they were. The confusion partially stemmed from the fact that of the hundreds of nebulae that had been mapped, many were not, in fact, the same thing. Some were simply clouds of gas, some were bunches of stars, and some were galaxies. Scientists who chose to believe that our galaxy wasn't alone could point to a nebula that was a galaxy and say, "Look, that clump of stars—it must be another island universe like our own." Others could point to a nebula that was a fuzzy cloud of gas and say, "Don't be ridiculous. There aren't any stars there, and if there are a few, they are still in our own galaxy." The problem came from assuming that all nebulae were the same. In the end, no one could prove whether we were a galaxy unto ourselves or whether there were other galaxies out there.

It was Hubble who first looked at a nebula and showed that the group of stars was far, far outside of our own. In 1923 he swung the Mount Wilson telescope up to look at a nebula in Andromeda. Not only did he spot stars, he also spotted a very special kind of star known as a Cepheid. Cepheids were, and still are, key to astronomers because they are the only kind of star to which we know how to measure the distance. Think about it: If you had to measure how many miles it is to our nearest star (other than our sun), Alpha Centauri, how would you manage it? There *is* a way to do it. As we move around the sun, our distances to Alpha Centauri and, say, Betelgeuse actually change. By comparing the way the stars appear to move in the heavens (and we're talking about changes impossible to see with the naked eye), we can use geometry to determine how far away we are from each. But these stars are close to us; go much farther and that movement isn't detectable. We can't use this method to measure stars that are very far away.

But Cepheids are different. Cepheids get brighter and darker over time, and in 1908, Henrietta Leavitt discovered that the speed with which they vary is directly related to how bright they are. Not how bright they appear to us, because, of course, things seem dimmer as they get farther away, but how bright they actually *are*. Once you know how bright a star is, and compare it to how bright we perceive it to be here on earth, you can figure out how far away it is. For example: imagine you see a flickering light off in the distance. If someone tells you it's a bonfire, you know it must be very far away. On the other hand, if it's a dim firefly, you know it's fairly close. And if you measured the brightness exactly, you'd be able to come up with some pretty exact measurements of just how far away it is. Using these methods, a scientist can tell you how distant any Cepheid is.

Hubble's Cepheid in that Andromeda cluster was very far away indeed—so distant that it clearly had to be outside our galaxy. The spiral nebulae were definitely other galaxies, claimed Hubble, and he kept studying them. (Much like relativity, the existence of other galaxies was accepted fairly readily. The idea that there might be other galaxies had been around for some time, after all, and everyone already agreed that Cepheids were an acceptable yardstick.)

Hubble began to study the "redshifts" of distant stars—that's the way their light appears redder than it actually is, since the galaxies are moving away from us. It's all part of the Doppler effect, the same thing that makes an ambulance siren sound higher as it approaches you and lower as it moves away—the very sound wave seems contorted to our ears due to the movement. Light is also a wave, and it appears changed to our eyes if it comes from a moving source. By studying the redshifts, Hubble measured just how fast all these galaxies were moving.

One day in 1929 he was looking at these galaxies and he realized that the motions weren't random. The farther away a galaxy was, the faster it moved. And he could predict the pattern: twice as far, twice as fast; three times as far, three times as fast. The theory is now called Hubble's Law, and it's crucial because these data were the first experimental evidence that the universe was indeed expanding.

If the universe was basically static, you wouldn't expect any kind of pattern. Galaxies would zoom in different directions, governed solely by the effects of gravity. Or let's say everything was moving steadily in one direction—like three people driving along a road at a steady twenty-five miles an hour—then you wouldn't expect to see redshifts at all, since nothing would be moving faster than anything else. But if all of space is expanding—the way a mound of bread dough rises in all directions—then it would match this pattern perfectly: the farther away something is, the faster it appears to be moving.

It was a fairly surprising result, but by the 1920s Hubble's career was well under way and his reputation was rock solid. He had burst into the limelight when he'd proved the existence of galaxies; and his science was known to be meticulous. Hubble had measured the speed and distance to so many galaxies that the evidence overwhelmingly supported an expanding universe. Confronted with such data, Einstein and the other holdouts finally agreed, showing that perhaps respecting the scientist involved is as crucial for believing new theories as anything else.

History, of course, isn't often as kind to a celebrity as contemporary acclaim may have been, and Hubble certainly has some modern detractors. There are those who correctly point out that he wasn't the first to suggest the relationship between distance and velocity, as well as those who claim he didn't truly understand the revolutionary nature of what he was proposing. Yet Hubble carried astronomy to reaches far outside our galaxy in a way no other astronomer had before him—and for that he deserves his fame.

This was about as far as anyone was destined to get for the next three decades. All the pieces were there, but no conformity, no final solution. It was agreed: the universe was expanding, and it was expanding according to the theory of relativity.

But that still left questions unanswered. Just because the universe was moving steadily outward didn't prove that it had all begun at a single point, or that it wasn't infinitely old. A river, after all, moves in one direction and yet is replenished. Did the universe perhaps maintain a steady state by constantly adding new material, the way rain constantly adds new water to a river's spring? This would be the debate that would dominate the next three decades of cosmology.

A Philosophical Debate

In the absence of hard evidence and observation, an interesting argument began to foment among physicists and philosophers at the time: Could cosmology as it stood even be considered a science? The discussion was loud and heated enough to take place in numerous public forums—including a series of opinion articles in 1932 in the *Times* of London between astronomers E. A. Milne and Sir James Jeans. The issue at hand was nothing short of whether theorists had any place in physics at all.

Jeans insisted that since the math behind relativity theory said that space was curved, then space was curved. The inductive reasoning worked, and so we must accept its conclusions as fact. Milne, on the other hand, said that science could be built only on what was observed. One could deduce principles logically from those experiments, certainly, but without some observable proof, he could never accept the curvature of space. In fact, Milne went so far as to say in a *Nature* article that since space was inherently unobservable, we should put an end to studying it altogether.[1]

Others—most outspokenly, Herbert Dingle—would soon take this even farther. Milne did allow for some initial axioms, which he believed to be self-evident, while Dingle claimed that no such thing was allowed. To do so was to reduce the very foundation of science to "invention." The only principles allowed, according to Dingle, were those that had a solid foundation in observation.

In essence the question was whether theory—the kind of theory we know today, the manipulation of math to create a system that makes sense—is an acceptable way to describe the universe. Or should anything so removed from the laboratory be kept in the realm of mathematics or philosophy? In the end a compromise was reached, as outlined by R. C. Tolman, who claimed that cosmology had room for both kinds of reasoning and could be considered a science that made use of inductive reasoning as well as observation.

In fact, as we shall see, cosmology has been largely driven by over-the-top, futuristic theories that are often ahead of the experiments needed to prove them. And philosophical arguments to accept theorists notwithstanding, this lack of observation dogged the study of cosmology through most of the twentieth century.

3

The Search for Proof

Despite the acceptance of an expanding universe, the field of cosmology was fairly slow to get off the ground. It just smacked too much of voodoo. For most of history, cosmology has been intimately entwined with religion. Physics, astronomy, biology—these fields had made a break from religion and philosophy in the 1600s. But questions about the very existence of the universe—how it worked, *why* it worked—were associated with theology, definitely not with science.

Very few people thought the ideas proposed could be tested, much less proven. Serious scientists were supposed to stick to what they could observe. In the case of astronomers, you looked at the stars and studied the images—you didn't presume to see into the past. In the 1920s there were only a few "truths" about the nature of the universe that were accepted and considered scientific: stars and galaxies were moving away from us; and Einstein's version of gravity had a great deal to do with why the universe was shaped as it was.

How to interpret that information was up for grabs. The facts didn't automatically point to a big bang theory, though some scientists were swayed in that direction. By the end of the 1960s the big bang theory was largely accepted. Why did the theory manage to stick? What convinced everyone?

The big bang theory really has only three major pieces of supporting evidence: the universe is expanding *à la* Hubble; the universe contains the right amount of helium and other atoms; and we've spotted what we believe to be the radiation left over from that initial explosion.

At a time when creation theories were considered anathema to "real" science, it's impressive that anyone thought to look for the second two bits of evidence. The names mentioned in the previous chapter—Friedmann, de Sitter, Lemaître—are some of the few who bothered to tackle the subject as a science per se. And none of them was exactly mainstream. The man who first came up with a scientific "test" for the big bang theory was even more of a maverick.

GAMOW THE GREGARIOUS

George Gamow was a talkative Russian who drove a motorcycle and who actually wrote books for a lay audience instead of just scientists. He was the loud scientist at conferences, holding court over drinks and doing impromptu magic tricks. You either loved him or you disparaged him—but either way, not too many people paid attention to his cosmology ideas.

Born in Odessa in 1904, Gamow studied physics at a time when the Marxist-Leninist philosophies of the Soviet Union were as involved with deciding what was "correct" science as they were with the experimental evidence. Gamow was taught, for example, that he could use Schrödinger's math to analyze quantum mechanics—but not Heisenberg's. The two kinds of math are absolutely equivalent (though the symbols and techniques used differ), and it's absurd to make any distinction between them. Religion was also getting short shrift in the USSR at the time, and even as a young boy, Gamow questioned Christian dogmas—he wrote in his autobiography that his first scientific experiment was to look at "transubstantiated" wine and bread under a microscope to see if it looked any different. As the political climate in the Soviet Union grew increasingly more restrictive, Gamow began to search for a way out, even trying to escape in a kayak across the Black Sea. Eventually Gamow and his wife left the country in 1933 when the USSR granted them visas to attend a physics conference in Brussels. They moved to the United States, where Gamow became a professor at George Washington University.

Gamow's uniqueness enabled him to emphasize the scientific nature of the big bang. Unlike many scientists, he refused to confine himself to just one discipline. He studied with Aleksander Friedmann in Russia, but he also studied particle physics with Ernest Rutherford—famed discoverer of the atomic nucleus—and quantum mechanics with Niels Bohr.* Gamow was one of the earliest nuclear scientists and was well respected for his understanding of the atom.

---

*After his work with the big bang, Gamow would go on to study biology as well.

Knowing particle physics as well as he did, Gamow understood that today's physics must be intimately linked with the creation of the universe. If the universe really did start from one single dense object in space, then whatever atoms exist today had to have been created in that first explosion—or else evolved from something so created. And so Gamow knew how to theorize the physics behind the big bang. The physics in the past had to be just right to create the physics of today.

Gamow wanted to understand why we have the precise amount of elements in the universe that we do. Astronomers knew that no matter which direction they pointed their telescopes, the universe seemed to be made of approximately the same stuff. Gamow suggested that the even distribution we see might have evolved from some basic, simpler stuff. His first version of the theory, published in a paper in 1942, suggested that perhaps the whole universe was once made of superdense, heavier-than-uranium elements that broke down into the myriad of lighter elements we see today.

Gamow was wrong. As it happens, he was wrong on the specifics a lot, but thinking to associate the stuff we see around us today with the stuff formed in the initial cosmic creation was so brilliant that it got him a lot closer to a workable theory than anyone had gotten before. It wasn't until 1946—in a *Physical Review* paper that may well have been the foundation of modern cosmology—that Gamow realized that the process more likely went the *other* way: the big bang created billions of lighter particles that coalesced into heavier ones.

Over the first half of the twentieth century, physicists had worked out a model of how atoms were made. Atoms are formed of smaller particles: protons, electrons, and neutrons. Different atoms have different properties—for example, hydrogen atoms form a light gas and gold atoms form a shiny metal—because they have different amounts of protons and neutrons in their nucleus. The very definition of an element is in how many protons and neutrons it holds.

Once bound up inside of an atom, protons and neutrons cling together by something called the strong force. The strong force holds them together incredibly tightly, but it only works at very short distances. It's like a sprinter or a relief pitcher: all strength, no staying power. And, of course, since protons are positively charged, if they're moving along freely outside an atom, they naturally repel each other. So, for two or more protons to be bound into a nucleus, they must get close enough for the strong force to overcome the basic repulsion. A question physicists asked in the 1930s was, How do those protons ever get that close?

They found the answer in the stars. The bulk of the work was done by a German-born American physicist, Hans Bethe, who would later win a Nobel Prize for his work. But Gamow and other scientists contributed to the discussion as well. The theory—which has been modified slightly over time but is still pretty much accepted to this day—was that deep in the heart of stars lies a scorching-hot furnace. The fuel keeping the flame alive is hydrogen. In the center of a star the heat is so intense and the pressure so massive that it forces the hydrogen atoms together—forces them so close that the strong force can catch the neutrons and protons and bind them together to form new, heavier elements.

Gamow ran with this idea. Interested in why we have the exact abundance of atoms we do in this universe, he hoped the stars themselves would give him the answer. They didn't.

The stars managed to create many elements but simply couldn't be responsible for all the elements we see in the universe today. (At the time, scientists erroneously thought the stars didn't make enough heavy elements. It was later discovered that the real problem was in creating light elements like helium and deuterium.) Gamow and Bethe both suspected that the bulk of those elements had to have been created even before the stars were.

This furnace idea, thought Gamow, was a good one, but it couldn't be confined to just the stars. Gamow's 1946 paper suggested that perhaps at some point in the past the whole universe was just a thick fog of sizzling neutrons. A very dense and small baby universe, these neutrons would be close enough to join together via the strong force. Once bound, some of the neutrons would turn into protons through a process known as beta decay. And *voilà!* The elements were formed. This universe was expanding, of course, and expanding rapidly according to Einstein's equations, but the origin of this neutron soup constituted a beginning, a creation for the universe. Gamow's description was truly the parent of today's big bang theory.

Soon Gamow joined forces with a graduate student, Ralph Alpher, and with a colleague, Robert Herman. Alpher's graduate thesis went farther than Gamow, studying how neutrons joined together and under what circumstances; he managed to create an even more precise picture of the universe's origin.

In the beginning, according to Alpher, there were only neutrons—small neutral particles today found in the nuclei of atoms. Many of these neutrons decayed over time into protons and electrons (positively and negatively charged, respectively), making a swirling, hot mix of particles Alpher named "ylem." (He found the word in Webster's. It was an obsolete word for the primordial goo from which all material formed.) At a certain point in the expansion, this boiling soup would cool down to the perfect temperature, about 6,000° Fahrenheit, for element formation—still hot enough to overcome the electric repulsion, not so hot that the particles moved too fast to connect. How or why the ylem was created from the initial explosion was unexplained.

Gamow put Bethe's name on the paper, too, even though Bethe had had nothing to do with it. Just because it sounded nice. He wanted to refer to the whole shebang as Alpher, Bethe, Gamow—or Alpha, Beta, Gamma, the first three letters of the Greek alphabet. Gamow didn't ask Bethe ahead of time, but Bethe was one of his friends, and he figured Bethe would appreciate the joke. The paper has always been referred to since as the Alpha, Beta, Gamma (or αβγ) paper. Gamow was so amused by this that he tried to get Herman to change his name to "Delter" for the fourth letter in the greek alphabet.

Alpher, Herman, and Gamow realized an important side effect of this cooling ylem. Initially the ylem would be so dense that no energy would be able to escape—photons zipping around would end up crashing into other particles and never escape the boiling mess. This energy, however, would be perfectly distributed, the way heat, over time, will become evenly distributed around a room. This is known as a state of thermal equilibrium, and since this primordial equilibrium had three hundred thousand years to equalize without ever letting a single drop of energy escape, it would have been one of the most perfectly even of all time. But as the ball of ylem continued to grow, losing density and heat, the mix would drop in temperature and become cold enough for radiation to escape. For the first time in the universe's history, light and energy would be able to travel free, streaming away from the initial cluster.

This escaping radiation would have been traveling ever since, with nothing to hinder it. And since it was so evenly distributed, the radiation would give off very specific spectrum lines. Like the barely perceptible static fuzz on a radio station, the background radiation would be underneath the light and energy moving around the universe. (Indeed, scientists today say that 1 percent of the static picked up on a home antenna is due to this radiation.) Gamow's team went so far as to predict a variety of temperatures for this radiation. They were, as it turned out, a little unsure on this point, predicting over the years anywhere from five to fifty degrees above absolute zero. The correct value is a little less than three degrees.

And then their research was forgotten. The idea of a creation of the universe was remembered, and embraced by some, but the specifics of Gamow's version and the possibility of background radiation were overlooked. Some scientists knew about it—Gamow certainly promoted it regularly—but it never quite intrigued anyone else enough to really focus on it.

One maverick does not necessarily make or break science. Even though Gamow offered a chance to find scientific "proof," no one followed up. This was partly because the technology of the day simply wasn't up to the task—separating that background noise from the brightness of the stars and galaxies requires extremely sensitive instruments (think about trying to hear the softest of white noise hums under an otherwise booming radio station). Many people assumed that technology would never be good enough to detect the background radiation—and Gamow's group certainly didn't make suggestions on how to do it. It's also possible that the team's predictions didn't sound firm enough, given the various temperatures they predicted. Or perhaps the theory's association with the gregarious and controversial Gamow was enough to keep it out of the spotlight.

Regardless, Gamow's suggestions were roundly ignored by the scientific community. The idea that the universe was expanding was fairly well accepted, but the thought of turning this moment-of-creation idea into an experimental, observable science simply didn't take root.

The big bang concept was not, however, ignored by religion. Gamow sent a copy of his theories to Pope Pius XII in 1951. The pope, who had recently endorsed the big bang theory, gave a speech that November saying that science had proven the existence of creation and a creator. This association with religion gave even more fuel to those who thought cosmology wasn't a serious scientific pursuit. Linking the universe's creation to Genesis gave the opponents fodder for dismissing the big bang as a retro theory as heinous as the medieval trend of harking back to Plato and Aristotle.

Challenging the Big Bang: Steady State Theory

With no room for experimentation or observation, these kinds of religious, philosophical, or just plain "gut" choices lay at the heart of whether one believed the big bang theory over any other. For many years, the decision of which cosmology to endorse was a qualitative one, simply a discussion of what made most sense to you—and so was a highly individualized choice. This is why a team of British scientists staunchly rejected the whole big bang idea. They trusted Hubble's data. They believed Einstein's relativity. They simply thought this fundamental first explosion was a ridiculous notion. The founders of the alternative theory, which came to be known as the steady state theory, were Fred Hoyle, Hermann Bondi, and Thomas Gold. The three men first met doing radar research during World War II. Hoyle was already an established professor at Cambridge; Bondi and Gold were young Jews newly escaped from Austria.

It was the early 1940s, and they formed a trio to discuss their foremost interest: astronomy. Gold was the quiet one; he listened and synthesized, often adding new directions or insights. Bondi had the sharpest math mind, and he was the one who scribbled out equations for the group. Hoyle was the dynamo whose mind skipped from idea to idea, often landing on a theory simply through intuition (much to the chagrin of Bondi, who would often rectify Hoyle's mathematical assumptions and misguided techniques, only to discover that his final results had nevertheless been correct).

After the war the three ended up at Cambridge together, where they continued their astronomy discussions and put together an alternative to Gamow's creation theories. For one thing, the Hubble law said the universe was about two billion years old. A nice, long time, to be sure, but about half as old as the earth. Scientists had already measured the age of certain rocks on our planet and found them to be closer to four billion years old. There was no way the earth could be older than the entire universe.

This little time-scale problem, however, was at most just an embarrassment. Scientists who embraced an evolving universe felt sure the problem would be corrected with better measurements, and even the British team knew there were ways around it. In the end their primary objections to the big bang theory were more subjective.

For one, they simply disliked the idea of a single "origin" to time and space. A moment of creation just didn't make sense to them. (Hoyle was particularly adamant about this: he claimed that believing in a creation of the universe was "primitive," harking back to the days of creation myths.) Second, that origin—or singularity, as it's called—seemed undescribable by standard physics. Einstein's laws broke down completely at the extremes of temperature and density described by the big bang. Hoyle's team wasn't eager to accept something that failed to conform to what scientists agreed they knew about the world. Last and perhaps most important, the evolving-universe idea couldn't be tested. With the exception of the background radiation—an idea of Gamow's that no one really remembered—the big bang theory wasn't even a scientific theory, since it was speculation without means of proof.

The "big bang idea" was the way in which Hoyle disparagingly referred to this type of creation theory, and that is how the modern theory got its name. He meant it mockingly, but "big bang" caught on, and eventually everyone referred to Gamow's ideas that way. (Hoyle belittled the big bang theory until he died in 2001. In the early 1990s, the magazine *Sky and Telescope* held a contest to rename the theory with something more appropriate. When no one could come up with a better name, Hoyle commented that it was just as well, since he still predicted that the whole concept would soon be overturned.)

In the late 1940s, Hoyle, Bondi, and Gold devised a new, testable theory. The universe, they stated, had always been as it is now. (This concept is called the "perfect cosmological principle," and it's a natural extension of Copernicus's insights nearly four centuries earlier: the earth is not in a special, central place in the solar system; the earth is also not in a special, central place in the universe; the earth isn't even in a special place in time.) Yet they couldn't deny that galaxies were moving away from each other.

So how do you maintain stasis—unchanging density—in an expanding universe? To keep the universe from having been denser in the past, or less dense in the future, Hoyle lit on a brand-new idea. Matter, he said, was continually being created. Out in the voids of space, between galaxies, new matter was being born to replace the ever-migrating stars. Not one creation at one moment in time, then, but constant creation—constant regrowth to keep the universe in a steady state.

The theory was testable because it made very specific predictions about the nature of the universe. Galaxies young and old would be equally scattered around the universe; as galaxies moved, they would decelerate at a specific rate; and the very shape of the universe was specified (while to this day, the big bang theory allows for a variety of shapes).

THE STEADY STATE VS. THE BIG BANG

Of course, the thing about a testable theory is that since it can be tested, it can be disproved. Constantly updated, occasionally altered to accommodate new data, the steady state theory lasted almost thirty years before experiments finally proved it dead. (Which is not to say that all its proponents now agree with the big bang theory; some still seek alternative explanations.)

The first observation to affect the theories was a reworking of Hubble's original data. Hubble had used the time-honored method of mapping blinking Cepheid stars to determine the distance to other galaxies. But in the early 1950s the techniques were deemed slightly off-kilter—the stars Hubble had analyzed weren't the "right kind" of Cepheid. New results showed that the universe was far larger than previously imagined. If the galaxies were farther away, then they must have been traveling for a longer period of time—the age of the universe could now be six billion to thirteen billion years old. The teeny problem of the earth being older than the universe was completely cleared up. Of course, there were estimates that some stars were more than fifteen billion years old, so there was still a time-scale problem, but the change in the age of the universe let everyone believe that there could soon be other adjustments in either the age of the universe or the age of the stars. The time-scale issue occasionally shows up to this day—the highest estimates for the age of the stars do not match up with the lowest estimates for the age of the universe—but most people attribute this to observational uncertainties instead of a fundamental problem with the big bang theory itself.

A variety of other challenges to the steady state theory soon arose. Two American astronomers, Joel Stebbins and Albert Whitford, showed that galaxies that were farther away were also younger. Since light from such faraway galaxies has taken so long to reach us, then the pictures we take are snapshots of what they looked like in the past. If *all* those galaxies were young in the past, then one assumes that our universe as a whole was younger then. It was growing up, evolving over time. In a steady state universe that had been around forever, there should be nothing special about looking back in time—you would expect to see young and old galaxies alike. Another blow to steady state theories came from California, where Allan Sandage was making observations that suggested the movement of the galaxies, the rate at which they slowed down over time, wasn't compatible with the steady state equations. This was good data, often quoted by supporters of the big bang—but enough contradictory information followed from subsequent experiments that the steady state theorists could, and did, dismiss it as controversial.

On the flip side, scientists studying star formation discovered that stars *could* make far more of the heavier elements than Gamow or Bethe realized. One of the main reasons why big bang proponents required that initial explosion was for the incredibly hot furnace that would create all the elements we currently see in the universe. But if the stars were cranking out those heavy elements, then the big bang wasn't necessary. Much like attacks on solid state theory, though, this was a setback but not a deathblow.

None of this controversy was specifically about the big bang theory as we envision it today. Current theories include a moment when the entire universe popped into being and expanded to create the modern universe. But the argument in the 1950s was between a steady state universe and an evolving one—evolving from some primordial state different from the current one. How we got to that initial "primordial state"—whether it was created out of nothing or created from a constantly oscillating universe growing, collapsing, and growing again, or something else entirely—could be shunted to the back of one's mind. If a point of creation or a mind-bogglingly dense atom containing the entire universe made you squeamish, then you could dismiss the details of how that ylem came to be as an unscientific problem never to be understood. And this gave the evolutionary universe theory an edge. While the steady state theory required constant creation, the creation of something from nothing, an evolutionary universe allowed one to ignore the whole messy question of how matter could ever arise from a vacuum.[1]

Today, of course, it's the exact opposite. One of the gut reactions one has to the big bang idea is "Well, what on earth existed *before?* How could something have come from nothing?" That unanswered question rubs many people the wrong way. In fact, that kind of philosophical issue was part of the steady state theorists' frustration. And yet the whole situation turned around on them. By using the word "creation" to describe the process of matter formation out between galaxies—a word they chose for scientific purposes, since it already was used to describe certain processes of particle interactions—they brought the issue to the forefront of everyone's mind.[2] Those who supported the evolutionary theories, on the other hand, could ignore the whole problem.

Again, as scientists picked a theory to support, these were the kind of intuitive decisions that came into play as much as anything concrete like "evidence." While most mainstream astronomers disliked the steady state theory, the actual experiments done in the 1950s gave one only enough information to confirm or deny the theory one wanted to believe. No evidence was definitive. There were people such as Allan Sandage, who said he thought of the steady state theory as basically dead from the moment it was conceived; and there were people such as Hoyle, who until he died didn't believe we have enough proof to accept the big bang.

This doesn't mean that observations didn't matter at all. Hoyle certainly knew that a theory was irrelevant if it didn't fit the data—observations lie at the heart of good science. In fact, this may have contributed to the downfall of his theories. While Gold and Bondi stuck close to the beauty and simplicity of the original steady state version, Hoyle continued to modify the theory to accommodate new facts. While in some ways this is exactly the way science is supposed to work—make a theory, test it, devise a new one if it fails—the continual adjustments made the theory look shoddy. The theory ended up scarred and constantly being patched up, like a leaky roof, in order to work. While no experiments had yet derailed it, the steady state theory was beginning to get banged up—and the evidence for an evolutionary theory would soon grow.

NEW DATA

One of the fascinating issues with any scientific cosmology model is that it predicts a "shape" for the universe. This is nothing so simple as wondering if the universe is a pyramid or a cube; this is a question about the very way space is *formed.* It's a four-dimensional issue and therefore not one most of us can really get our brains around, but you can get a basic idea by thinking about the problem in three dimensions. As you walk around on the earth, there's really no reason to think you're walking on something round. It appears flat. Knowing as we do that our eyes can play tricks on us, making the world seem flat when we're really walking around on a sphere, you can imagine that if somehow we were walking on the inside of a gently sloping bowl, instead of the outside, that might appear to be flat, too. If the curve is gentle enough, the tricks to figuring out what kind of shape you're walking on are not going to depend on what you can see. One option would be to walk as far as you can in one direction and see where you end up. If you come back to where you started, you know you're on a convex surface. Of course, not ending up where you started doesn't tell you much of anything—the earth might be flat, it might be a gigantic bowl, or maybe you just haven't walked far enough yet.

None of the ancients used the walk-around technique to figure out that the earth was a globe. It's only recently that airplanes or even ships circumnavigated the world. Ancient observers used different techniques, like the way the masts of a ship were the first to appear on a horizon, or the way the earth casts a round shadow on the moon, to measure the shape of their world.

But with the universe itself, we have yet to figure out what would happen if a spaceship set off in one direction for billions of light-years. Would it go straight? Curve ever farther away, as if in a bowl or a saddle? Curve and eventually come back to the beginning? Einstein's laws of relativity could give a very specific answer—since it's gravity that causes space to twist and bend in this way—but those laws depend on how much matter exists in the universe and how it's spaced out, information we can make good estimates of but can't guarantee. And so scientists come up with tricks to try to measure the exact curvature of the universe.

The steady state theory predicts an absolutely flat universe, while the big bang theory anticipates a universe that might be flat or curved (though most modern-day cosmologists believe it's most likely flat—more on that in chapter 4). Looking out over a flat universe, certain things should be evenly distributed. (If they were lining the inside of a big bowl, for example, the distributions would be distorted.) Add that the steady state theory insisted that new stars and galaxies were constantly being created throughout space, and there's all the more reason to think that the universe should be pretty much the same whichever way you look, no matter how far away you look.

One of the things scientists could observe was the average energy at any given moment being pumped out of stars across a given space (technically known as the "flux density"). If we live in an even, flat universe, as steady-state theory predicted, then a very specific number of stars should have a very specific range of flux densities. Measuring this outpouring of energy became easier with the invention of radio astronomy.

Light rays are, of course, made up of numerous wavelengths of light—the different wavelengths correspond to different colors. But there are many more wavelengths than just the bits our eyes can see, and anything that radiates gives off far more invisible wavelengths than visible ones. Measuring the invisible radiation, such as radio waves, streaming from stars turns out to be as informative as measuring the basic light we can see with our eyes.

In 1955 Martin Ryle, another British scientist at Cambridge University, used radio telescopes to map more than nineteen hundred radio sources outside the Milky Way. The flux densities didn't correspond to a perfectly homogenous universe at all. There was a flare of excitement—if correct, the evidence would rule out the steady state theory once and for all. Score one for evolutionary cosmology.

But the victory wasn't to last long. Other scientists failed to confirm the results, making this just one more piece of information that added fuel to the fire but proved nothing.

Until 1961. With updated tools and more rigorous work, Ryle and his colleagues came up with new data. This time the work was solid and hard to criticize, and other scientists soon repeated the experiments. The range of flux densities just couldn't be coming from a flat, even universe. If you believed Ryle's data—and they were hard to deny—then the basic steady state theory couldn't be correct.

More crucial data soon came in, in the form of quasars. Quasar stands for quasi-stellar object, and when quasars were first discovered in 1963, they were a complete mystery—the size of a star, they burned as brightly as a galaxy. It was completely unclear what these gigantic, brilliant bodies were made of or how they could produce so much light, but it did seem likely that they were young objects—perhaps associated with newly forming galaxies. Their redshifts were so high that they appeared to be traveling very fast and to be very far away. Lots of young galaxies far away jibed with the big bang theory—not the steady state theory, which would have had evenly distributed galaxies, young and old, everywhere.

This was all strong evidence against the steady state theory, but nothing that necessarily proved the evolutionary theories.

Ironically, one of the main supports for the big bang theory would be work done by Fred Hoyle himself. Hoyle, along with Willy Fowler and Geoffrey and Margaret Burbridge, compared the amount of helium in the universe today with the amount that could be created in either of the two cosmology theories. The problem with the stars and just how many particles they could produce had reared its head again. It had already been established that the big bang wasn't needed to create all the heavy elements—the stars could take care of that just fine on their own. But all the helium, and other light elements such as deuterium and lithium, just couldn't be made by the stars. There is so much helium in the universe, and the stars pump out such relatively tiny amounts of new elements from their hydrogen-burning cores, that helium has to come from a furnace larger and hotter than your average star. (There are not, on the other hand, such huge amounts of deuterium, but stars destroy deuterium as fast as they make it. The fact that deuterium exists at all means it has to come from somewhere else.) The four astronomers wrote a book-length article now known as the B^2FH paper, after the last names of its authors, claiming that these light elements must have been made in some fiery source other than stars.

Hoyle, of course, pushed for an interpretation wherein large objects, perhaps black holes, played a larger role in atom production than was thought in a nevertheless steady universe. But most astrophysicists chose to believe a simpler solution: Helium and deuterium were cooked up at an earlier time, in a fantastically hot baby universe—a legacy of the big bang.

Final count, then, in the early 1960s: The evidence was building up against the steady state theory, and two pieces of evidence—the receding galaxies and the amount of helium—could fit into either theory but made a bit more sense in the big bang. And on the nonscientific side but not to be discounted, the evolutionary universe had captured the imagination of most scientists in a way that the steady state theory simply failed to do.

One last piece of evidence was all that was needed to end the debate once and for all.

THE FINAL BLOW

It's easy to make a big to-do about the fight between the steady state theorists and the big bang theorists—as if the only thing standing in the way of accepting the big bang theory was its steady state rival. Indeed, at times it was a fairly heated controversy, but if one looks closely, the noise was being made by only a few scientists: a handful of physicists in the United Kingdom for steady state, a slightly larger group of physicists, both European and American, for the big bang.

Most scientists continued to view the whole subject as too nebulous to get into.

Two such scientists were Arno Penzias and Robert Wilson. Wilson kind of liked the steady state theory, if pressed to pick a contender; Penzias didn't think much of cosmology either way. Yet, by an accident of fate, the two were destined to win the Nobel Prize for discovering the best "proof" available for the big bang theory. They were the first to hear that distant hum of the background radiation predicted decades earlier by George Gamow.

Penzias and Wilson weren't looking for the radiation at all. They were using an abandoned antenna at Bell Labs that was originally used for satellite communication. Penzias was the more talkative of the two. Born in Nazi Germany, he had come to the United States as a child. Wilson had grown up on the West Coast and tended to be more reserved. For both, the Bell Labs job was their first professional opportunity after graduate school—and they were the only two radio astronomers at Bell. The company had a long history of investing in basic research, and when their antenna—a giant twenty-foot-long cone lying on its side that was set up to receive satellite signals—became outdated, the labs thought to use it for radio astronomy. The antenna was well suited for the task, since its number-one job in its previous life had been to listen for signals at a very specific wavelength in the midst of all the radiation streaming in from the sky.

Bell Labs hired Penzias and Wilson as radio astronomers and gave them free rein to modify the antenna as necessary. The pair worked together well, and quickly upgraded the antenna to perform their first tests.

In June 1964, the two Bell scientists pointed their antenna at the sky to measure any radiation coming in at a 7.35-centimeter wavelength. This was merely a preview, a test to see if the antenna was calibrated. This wavelength just happened to be what the telescope had been set up for previously, and it seemed as good a wavelength to test out as any, since the Milky Way emits nothing in this spectrum. The scientists expected to hear nothing. If they got absolute silence, they would know there was no extra noise and they had calibrated their antenna perfectly.

So it was quite a shock when the antenna picked up a clear signal at 3.5 degrees above absolute zero (that's about −270°C). This wasn't what was supposed to happen at all. First, Penzias and Wilson went over all their adjustments on the antenna. They couldn't find a problem. Then they started pointing the telescope in different directions. Could the radiation be coming from some local, man-made source? But no, pointing the contraption at various nearby cities didn't affect the constant 3.5-degree reading at all. In fact, now that they thought about it, the reading remained steady no matter where they pointed the antenna. This radiation didn't seem to be coming from any specific star or from the galaxy. It was evenly distributed. And that made them all the more convinced it couldn't be coming from the heavens; it must be some problem of the antenna itself.

So they searched for alternate explanations—for ten months. Their best and biggest hope came in the form of the antipigeon campaign. Nesting inside the cone were a pair of pigeons, and Penzias and Wilson did everything they could to get rid of them and all the residue they left behind. (A paper the pair wrote euphemistically referred to the bird droppings as a "white dielectric material.") After completely ridding themselves of the pigeon problem—a job that involved removing the birds, only to have them find their way back—and cleaning the antenna from top to bottom, the two men discovered that absolutely nothing had changed. The noise remained. They were stumped.

Finally, Penzias mentioned to a colleague, Bernie Burke, the problems they were having with their giant antenna. Burke gave Penzias the key to the puzzle. He'd just heard from a friend of his, Ken Turner, who'd gone to a lecture by Jim Peebles. Peebles had reported that at Princeton—a mere thirty miles away from Penzias and Wilson—astronomer Bob Dicke and his graduate students Dave Wilkinson and Peter Roll were building an antenna to search for the background radiation of the universe, radiation that they predicted should be right around the temperature Penzias and Wilson had spotted.

A quick call to Princeton, and Penzias confirmed what he'd been told. Without knowing about Gamow's work, Dicke had rediscovered the idea that the universe had started in an initial fireball. As it happens, Dicke hadn't been searching to confirm the big bang theory per se. Much like the steady state theorists, he didn't like the idea of a single beginning to all time. Instead, he theorized that the universe had existed for all eternity but was constantly expanding and collapsing again. Dicke had to deal with the same element issue that the big bang and steady state theorists did, but his problem was slightly different. While those theories wrestled with how to create the exact abundance of elements we see today, Dicke's eternal universe would have too *many* heavy elements—since previous universes would leave residual matter. As the universe expanded—something that Dicke believed it had done over and over and over—there must have been a fireball to destroy some of the heavy elements left over from the last go-round. And to prove his fireball, Dicke set out to search for vestiges of that initial radiation. Unlike when Gamow predicted it, however, the tools now existed to find it. Dicke, Wilkinson, and Roll were just about to start their experiments on an antenna they had built from scratch when they were scooped by the Bell Labs team. Both teams wrote papers for *Astrophysical Journal Letters*—Penzias and Wilson describing their results, and the Princeton group offering the interpretation that this was the result of a great, primordial fireball.

Reactions were mixed. Many people laughed. Here was one sole data point in a quasi-scientific field; for those used to concrete experiments in the lab, it wasn't exactly overwhelming "proof." On one side, finally vindicated, was George Gamow. But he wasn't particularly happy. Nowhere did anyone mention his work with Alpher and Herman. Nobody bothered to report that they had predicted this radiation decades earlier. It could be, and has been, argued that Dicke and his colleagues should have done more research in their own field to study the predictions of the scientists before them. However, the fact that they didn't do this research makes sense given the nature of the field at the time. Cosmology was *not* an established science yet. Many people in the field saw themselves as basically working alone, on entirely new material. There was little collaboration, no comparing of ideas among universities, and no cosmology journals to research.

That would soon change. Within a year of the Bell team's work, Roll and Wilkinson had confirmed the results. Experiments have always been what defines science. If you're not out there observing and testing your hypothesis, then that hypothesis fits squarely in the realm of philosophy or, perhaps, religion. So finding a way to truly test the big bang theory lent the field its first legitimacy. In 1978 Penzias and Wilson received the Nobel Prize for their work, an impressive award for a result they'd never set out to find. Wilson said that he didn't realize what an important discovery they'd made until he read about it in the *New York Times*.

The work is generally heralded as the first real proof to support the big bang (and to this day, the existence of the cosmic microwave background is one of the most important pieces of evidence). It was now generally agreed that the universe evolved over time, starting from some initial form of radiation and particles. The big bang theory had finally entered the world of "real" science.

Interlude: Popular Reactions

While scientists do all the molding of a theory, scientific ideas always filter down—almost as "fashion"—to the rest of the populace. In turn-of-the-twentieth-century France, for example, just after X rays were discovered, the chic parties all boasted X-ray machines to take pictures of the guests' hands. A few years later, with the discovery of radioactivity, uranium became the must-have ingredient for everything from hair tonic to Day-Glo paint. (This particular fashion not surprisingly lost its panache when it began to result in untimely deaths.)

Science clearly inflames society's imagination. What's less obvious is that these kinds of reactions can add support for any given theory. But one need only look at the insistence of even Galileo and Copernicus, who held that planets moved in perfect circles simply because that was what they'd always been taught, to understand how intrinsic the popular viewpoint can be. By rooting a theory in nonscientific language for the layperson, absorbing it into the literature of the time, and inviting religious and philosophical inquiry, the popular reaction to a scientific endeavor can do much to cement it into "fact." Getting a scientific theory seeded into the public consciousness can in its own way contribute to a theory taking hold as much as anything else. And the big bang theory, of course, describing nothing less than the creation of the universe, has captured the imagination even more than most.

RELIGIOUS REACTIONS

The simplest religious reaction is the fundamentalist, absolute reaction. Those who believe dogmatically, literally, in their religion's version of how the universe came to be will always trust in their faith more than in the scientific method. Many very vocal groups take serious issue with the science behind the big bang. In the fundamentalist Christian case, for example, the big bang contradicts the Genesis story and insists on a universal time frame millions of times longer than the under-six-thousand-year time span implied by the Bible. These contradictions with the Christian doctrine will never be accepted.

When discussing religion and science, much press is devoted to this kind of stark incompatibility—as if they were extreme opposites, as if the only possibility for someone who is religious is to unquestioningly deny experimental observation, and the only possibility for scientists is to starkly accept only what they can see with their eyes, denying the near-religious beauty, near-mystical quality the universe can hold. In fact, the field of cosmology has been populated with the devoutly religious: Lemaître was a Belgian monk, Kepler studied for the clergy, and Galileo cared deeply for the Catholic Church, to name a few. For such people there has never been a contradiction between their doctrine and their science. It is far too easy to assume that religion and science are always at odds with each other.

Science and religion more often either comfortably ignore each other or complement each other in subtle ways. The initial widespread acceptance of the big bang theory is a fine example. Humans had been conditioned by religion for millennia to believe in a moment of creation for the universe. A scientific version of what they'd always thought was true anyway certainly appealed to the masses. In fact, the initial response from some religious leaders was overwhelmingly positive: In 1951 the pope gave a speech saying that science had finally "proven" creation and the presence of a creator. (Interestingly enough, the intellectuals of the Catholic Church are usually given much less credit than they deserve for embracing new scientific ideas. Galileo's trial is often heralded as the supreme example of their closed-mindedness against science, but the trial was more political, more about the threat to the church's absolute power when it was already sustaining attacks from Lutherans, than about defending traditional Christian doctrine.) While calmer pontifical heads soon downplayed the pope's outspoken viewpoint, pointing out that the church didn't need confirmation from scientists, nor vice versa, the almost instinctive desire to link the big bang with Genesis was an obvious step for many.

On the flip side, many supporters of the steady state theory—most outspokenly Fred Hoyle—specifically liked the fact that the theory didn't jibe with the classic, biblical version of events. While its lack of justification for Genesis was never held up as the most important thing about steady state, it was an attractive side point. But in a Western culture that is fairly comfortable with a creation story, the atheistic association of steady state wasn't necessarily going to endear it to the population at large—another subtle reason why many didn't embrace the theory on a gut level. (Not everyone saw steady state theory as devoid of the divine, however; astronomer Bernard Lovell, for example, a devout Christian, believed that the continuous creation of matter required by the theory was as good a sign of God's activity as any.[1])

Making associations between cosmology and religion is not just a habit of the religious. Many who trust the scientific method whole heartedly nevertheless have an awe for the beauty of the cosmos and its origins, which borders on the unexplainable or the mystical. The very complexity of the universe suggests to them that a divine hand—even if not a Judaeo-Christian God sitting with a mighty scepter on a throne—must have been behind the greatness.

This kind of "beauty must be divine" concept is an instinctual one for many people, but some scholars have delved into it on a rational, philosophical level. Our current model of the universe doesn't allow for the geocentric belief that a god created a world with us at the very center, but the religious concept of "for my sake was the world created" doesn't need to be scrapped. This type of universe, this type of solar system, this type of planet, are the only ones in which we could live—and that means we are very special indeed. One could argue that the universe was created with intelligent life in mind.

This is known as the "argument from fine tuning," and it essentially goes like this: The incredible precision needed in the first moments of the universe to create intelligent life is awe-inspiring. If the expansion rate of the early universe had been different by even one part in a million, galaxies would never have formed. If the force of gravity had been off by that amount, stars wouldn't survive long enough to create hospitable planets.[2] If the strength of the nuclear weak force were different, all hydrogen in the universe would have been burned into helium, and there would be no water. If the nuclear strong force had been off by as much as 1 percent, nowhere near enough carbon for life would have formed.[3] Addressing this kind of accuracy, British astronomer royal Martin Rees recently wrote a book titled *Just Six Numbers,* naming the six numbers that were crucial for life to have formed. They are the "strong force," which binds all protons and neutrons together in atoms; N, the number of times greater that force is than gravity; Ω, the shape of the universe; λ, the cosmological constant; Q, the size of early fluctuations in density that seeded galaxy formation; and D, the number of dimensions we experience. Something at the time of the big bang made these numbers just the correct value for humans to be born. Since the values of the numbers don't seem to depend on each other in any way, it's all the more incredible that they turned out just perfectly. If, for example, the size of one determined the size of all the others, then their precision wouldn't be so awe-inspiring.

Scientists do attempt to explain how a random quantum energy fluctuation could have lucked out and created this precision, but the fact is that no proven solutions have been found. In the absence of such answers, there can really be only one conclusion, say supporters of the argument from fine tuning: some creator must have made it that way. Some creator initiated, or helped adjust, a big bang that would lead to the existence of intelligent life.

In this way, the fact that there was a big bang at all, a big bang so precisely tuned to grow humans, is used to prove that there must be a God. (Interestingly, Rees himself doesn't see his numbers as proof of God. In fact, the random values of so many of the numbers—λ being just a bit above zero, for example—suggests that they must have been made by chance or as a consequence of unknown physics. A god would have made the numbers a good deal simpler.)

And, more importantly for us in this discussion about why the big bang was accepted, this natural jibing with the religious beliefs in our culture offers the populace more gut-level reasons to subscribe to the big bang theory. In a world where many people believe in the divine, the wonder inspired by the big bang makes it easy to accept.

PUBLICATIONS

Another place where religion meets cosmology is in the realm of popular literature. A whole slew of books with titles such as *The Tao of Physics* and *Genesis and the Big Bang* attempt to assimilate scripture (Buddhist scripture in the first example, Jewish in the second—but there are many other similar books, discussing other religions) into big bang cosmology. The manner and styles differ dramatically among the books and the various scriptures involved, but the basic purpose is the same: to seek echoes of the religious origins of the universe in the scientific versions. Dwelling on light and energy, many of the writings from Eastern philosophies sound almost as if they were precursors to modern theories. Discussing the Western tradition, *Genesis and the Big Bang* argues that both versions of creation describe the same events—they simply use vastly different languages.

From the scientific perspective, saying that any religion predicted big bang cosmology, or had somehow "known all along" how the universe was created, is fairly far-fetched. At most, the consistency comes from our human addiction to certain metaphors—religions and the sciences both have relied on light, energy, and creation out of nothing to describe their beliefs. But when it comes to accepting a new scientific theory, that consistency is important. At its core, science is a set of mathematical equations combined with an enormous amount of observation. For most brains, this isn't enough to create a fully formed picture of how the universe works. Scientists must interpret the data, must use words—inherently much less precise than math—to describe what they see. They must create a verbal theory to explain their experiments. Naturally, the metaphors used to visualize a new theory will be drawn from the culture at hand; surrounded by such familiar metaphors, one can see why the culture would readily accept it.

Again, religion, too often seen as the antithesis of science, comes through in a subtle way to support the big bang.

PHILOSOPHICAL REACTIONS

Many of the deliberations about whether the big bang assumes a creator are done by philosophers as much as by theologians, but philosophy tackles other areas as well. Just about anything that can't be tackled from an experimental standpoint is labeled "metaphysics," and therefore all the "big questions," such as why we are here at all, get shunted off into the realm of philosophy. (This is not to say that physicists don't tackle the subjects, only that they do so in an extrascientific manner.)

The main "why we are here" argument is known as the anthropic principle. The idea was first outlined by Brandon Carter in 1974. At its most basic, the principle is very benign: The universe must be a universe that can support intelligent life, otherwise we (i.e., intelligent life) couldn't possibly be here to wonder about it. This seems fairly obvious and perhaps even trivial, like saying of course the room I'm sitting in is big enough for a human to sit in, or else I wouldn't be sitting in it. But the principle has caused a fair amount of controversy.

For one thing, many people find it an unsatisfying answer. With the universe so finely tuned for human life, it seems highly unlikely that the effects of the big bang, with their precise values for the four forces, could have happened completely randomly. It's been said that the existence of a human-compatible universe happening purely by chance is as likely as a tornado tearing through a junkyard and creating a Boeing 747. To sweep such a coincidence under the rug, to dismiss it with a philosophical shrug of "Well, it just *has* to be that way," goes against the grain of most curious humans. Philosopher John Leslie put it this way: A man in front of a firing squad of one hundred riflemen is going to be pretty surprised if every bullet misses him. Sure, he could say to himself, "Of course they all missed; that makes perfect sense, otherwise I wouldn't be here to wonder why they all missed." But anyone in his or her right mind is going to want to know how such an unlikely event occurred.[4] And so the anthropic principle tends to fail as a reasonable explanation for why we are here.

Another version of the anthropic principle causes even more problems. The "strong" version of the principle goes so far as to say that the universe is endowed with a certain goal of creating intelligent life. No universe could exist without something in it to observe it, and so consequently the universe quite simply was required to be one with observers in it. One simply can't exist without the other. This idea is fairly hard to swallow: Why shouldn't a universe exist without intelligent life? But certainly as physicists delve further and further into quantum mechanics, it seems clear that much that occurs on the level of particle physics is strangely tied up with what an observer witnesses. Perhaps the idea that a universe requires a watcher isn't so strange after all.

Scientists often dismiss the anthropic principle completely. They want, after all, a scientific explanation for why the universe exists just as it is. They seek reasons why the four forces of nature were required to have the exact values they do. Or, alternatively, reasons for why our universe isn't as special as we think, one common explanation being that our universe is merely one of many, most without intelligent life. In a multiverse of so many universes, it becomes much more likely that one turned out to evolve as ours did.

On the Differences among Science, Religion, and Philosophy

For most of history, the search to understand our origins has been adamantly nonscientific. While the field has now become part of physics, it is, and probably always will be, entwined with religion and philosophy. It's crucial to realize that the kinds of arguments used in all three areas are going to be vastly different. Science's explanations are meant to be an accurate representation of some outside reality. Religion and philosophy are much more abstract, using logic to describe the universe and allowing for a fair amount of individual interpretation.

But those differences do not mean that the three fields don't, or shouldn't, communicate with each other at all. Religion and philosophy are naturally equipped to tackle questions that science can't solve, and on a subtle level, personal philosophies and religions will inherently affect the popular reaction to any new cosmological theory.

PART II

How Good a Theory Is It?

4

Scientific Reactions

Most scientists today believe emphatically that the big bang happened. Not that it's merely pretty math, or that there's a good chance, or that it's the best theory we've come up with so far . . . but that some ten billion to twenty billion years ago, our entire universe was really and truly compressed to tiny dimensions, which then swelled up, creating all of space and time.

It's hard to remember, given how accepted the big bang theory has become, that the first real proof was discovered a mere forty years ago. But the groundwork had been laid for decades—the ideas had been batted around and people were familiar with them. They didn't need much convincing once the necessary observations came in. Once the cosmic background radiation had been found by the Bell team and confirmed by the Princeton team, the astrophysics community went almost overnight from pooh-poohing cosmology as fantastical to embracing it as a bona fide, legitimate field.

Now that it was legitimate, the time had come to hammer out the details. And it's in the details where we soundly leave the world of nice, reproducible science. Scientists can't make repeated universes to see how they'll come out the next time around. They can't watch a star form and see how it turns out. They can't leave the solar system to look up close at anything other than the earth and some of the nearby planets. Experimental information has been hard to come by.

It was said in the 1960s that there were two facts about the universe: It was expanding, and the sky was dark at night (which basically means the observable universe must be finite—an infinite universe with infinite stars would be immensely bright around the clock). Some added an extra "half" fact: An initial explosion, the big bang, truly had occurred. Whether you were in the two or the two and a half camp, there wasn't a lot to go on. One could make up thousands of theories to accommodate these facts, and the standard big bang picture was merely one.

Almost no facts, but many, many questions. What triggered the big bang? What materials were created in that explosion? How did large structures such as stars and galaxies form? The only way to tackle these was to jump into the territory of theory and math. Many of the "add-ons" that have been tacked onto the big bang theory and that together make up modern cosmology are embraced because the theories complement each other and in certain ways correspond to what we see when we point a telescope to the sky. But truly testing them is almost impossible with contemporary tools.

This paints a fairly bleak picture. But there are ways to formulate a reasonable theory. The modern universe—obviously—exists in a very specific configuration. We have specific kinds of stars and specific kinds of galaxies; we have abundant hydrogen and we have intelligent life; and these current manifestations were precisely determined by the minutest of effects in the early universe.

Take, for example, the whole shape-of-space concept. While scientists have yet to determine whether the universe curves like a sphere or like a bowl or is perfectly flat (for more discussions on the shape of space see chapter 3), they do know it's at least fairly close to flat. A flat universe is one in which a number known as Ω (omega) equals 1. (A variety of factors can contribute to the size of Ω, but in the simplest models Ω is 1 when the density of the universe equals the "critical" density of the universe—the exact density at which the universe would expand forever, constantly slowing down but never coming to a complete halt.) If Ω is less than 1, our universe is "open"—shaped more like a saddle. If Ω is greater than 1, our universe is "closed"—shaped more like a sphere, wrapped around on itself. Using a variety of techniques, astronomers say that the modern-day value of Ω could be from 0.1 to 2. That's not precise enough to determine the exact shape of space, but it *is* precise enough to determine what Ω was a second after the big bang.

Since the universe has expanded so dramatically since that first second, any tiny anomalies at that point would have substantially increased over billions of years, like bank interest in a *long*-term investment. Going backward in time, if the universe is at $\Omega = 0.1$ today, it was a lot closer to 1 in the past.

In fact, it would have to have been 0.99999999999999. On the flip side, if $\Omega = 2$ today, it was *also* a lot closer to 1 in the past—specifically, it would have been 1.00000000000001.[1] The difference between those numbers is so small that were we not dealing with such precise calculations, anyone in his or her right mind would simply round them both off to 1. But that precision allows scientists to do some very real calculations about what the shape and composition of the entire universe was fifteen billion years ago, a second after all of space was created.

Observing the world today, scientists believe, allows one to "know" what the universe was like at the beginning of time, and so a very valid constraint is put on all theories about the big bang: they have to create a universe that looks like the one we live in today. Earlier scientists had done the first bits of work along this line—Gamow's first big bang model was specifically created to account for all the elements we see today—but with modern data, the amount of information about our current world has dramatically increased. The theories about the initial universe, the amount of material in it, the strength of the forces—all of these must be precisely calibrated to evolve into the amount and types of stars and galaxies we see today, the exact amount of the elements, a planet capable of sustaining life, a universe that is tantalizingly close to perfectly flat.

A lot of information to include, to be sure. Luckily, the timing was just right: supercomputers were coming into their own. These fastest and largest of computers could take an initial singularity, with very specific attributes appropriate for any particular version of the big bang, and then let this cyberuniverse evolve over time. If it doesn't come out looking like reality, then it's back to the drawing board.

This was the toolbox used to tackle the first questions raised by the big bang.

THE HORIZON PROBLEM

The universe is smooth. Hypersmooth. Sure, there are stars and galaxies, but if you could step back to some hypothetical vantage point and take a good look at all of space, you'd barely notice that there were these bumps of matter. It would be like looking down on the vast Pacific Ocean from an airplane—even though you know that ripples and waves exist, it would appear glassily even. The universe is that smooth.

So the first task of any scientist tweaking the big bang was to come up with an initial singularity that would create the smooth universe we see today. That's easy enough: the initial universe simply had to be extra smooth, too, since the tiniest anomalies would have been amplified over time into great, big distortions, creating a universe that today would look more like an ocean with tsunamis everywhere. The background radiation found in the early 1960s confirms this, showing a perfectly heated, even-temperature spectrum no matter which way you looked.* But this poses a problem.

*Technically, the temperature radiation is slightly different in one direction than in the other, but this is attributed to the fact that our galaxy is moving with respect to that radiation and not to the radiation itself.

The radiation is coming from two different directions, from areas of the universe so far away that they have no reason to be similar. In the standard big bang model there's no way, even in the first moments of the universe, that they were close enough together for a long enough time to reach the exact same temperature.

Think of it on a smaller scale. You pour two measuring cups of hot chocolate into a mug—that whole mug is going to stay the same temperature. All the hot chocolate molecules are in contact with each other, so they exchange heat back and forth, keeping the heat across the whole mug the same. But now pour a cup's worth apiece into two different mugs and move them across the room from each other. Information about the temperature will not flow from one mug to the other. If one mug is in a drafty part of the house and another is next to a radiator, you expect them to end up at different temperatures. In fact, if they remained at the exact same temperature, you'd be surprised. In the same way, measuring the temperature of the universe in two different directions and coming up with identical measurements shows that something unique is going on.

The explanation is fairly straightforward: the universe was once much, much smaller and there was time for everything to reach the same temperature, just as the entire pot of hot chocolate on the stove reached the same temperature before being poured out. Separated across billions of light-years, different parts of the universe nevertheless evolved in similar situations and retained similar heat spectra.

All of this makes a certain sense. At one second after the big bang, the universe was incredibly smooth and evenly heated. Consequently, our universe today is incredibly smooth and evenly heated. But how did it get that way from the beginning? In the basic big bang model, the universe was indeed minuscule in those first moments, but it stayed that way for only the barest fraction of a second. The universe, small even though it was, didn't have enough time to reach a uniform temperature. So somehow it must have been smooth from the get-go, instead of having a chance to grow smooth with time.

The question is why. This is known as the horizon problem (since information from the horizons in two different directions are identical). What events would lead to such a situation? Scientists can envision an unlimited number of beginnings to the world—plenty that would provide a fairly complex state. Why would the laws of physics favor a smooth one?

Asking "Why?" is, of course, when scientists get into sketchy territory. At a certain point the only answer to "Why?" is "Just because"—the point at which you can only find answers that make sense to you based on faith or opinion. Initially the question "Why so smooth?" smacked of that kind of problem. It could have been chance, it could have been a creator, it could have been anything.

A young postdoctoral physics student named Alan Guth had just this attitude toward the first microseconds of the universe's existence. These unsolvable problems, with their heavy dose of suppositions and philosophical musings, simply didn't satisfy him intellectually. He wanted science that could be tested, that made sense (although his focus at the time on magnetic monopoles—hypothetical particles that have never been observed—may seem more obscure than the big bang to many). But well-respected scientists around him were beginning to turn toward cosmology, and soon Guth would be caught up in the excitement.

One of his friends, Henry Tye, was fascinated by Grand Unified Theories. These are theories that hypothesized that in the intense heats and densities of the first moments of the universe, all the forces were essentially the same. As the universe expanded and cooled, the forces separated into the very distinct ones we experience today, such as gravity and electromagnetism. Grand Unified Theories or GUTs were yet another piece to add to one's big bang model. If you were a theorist who worked in the field, this might have comforted you with more facts to be examined; if you were Alan Guth, you thought you'd just added more thoroughly unsolvable questions to an already absurd puzzle.

But Henry Tye pushed him. A GUT combined with the big bang should predict the amount of magnetic monopoles in the universe. Since Guth understood monopoles—and knew that none had been observed so there must, in fact, be very few and be very hard to detect—Tye wanted him to do the calculations. At first Guth resisted, but as the ideas percolated in his brain, he slowly began to get intrigued. In addition, the climate of the physics community was changing. Suddenly scientists—*respected* scientists—were jumping into the cosmology game. It was no longer a field solely for the fringe element, but was becoming mainstream. And Guth thought he might be able to contribute.

Guth began doing the math to explain what might have happened during the first instants after the creation of the universe. He focused on an initial state of energy known as the Higgs Field, which, as it cooled and expanded, would essentially "freeze" into particles of all kinds, in much the way that liquid water can freeze into chunks of ice. Early versions of the equations had the field freezing into as many monopoles as there are protons and neutrons—an unlikely possibility in a universe where everything we see today is made of protons and neutrons and there's not a monopole to be found. So Guth kept molding the math.

He realized that in certain circumstances, the Higgs Field would create what's known as a false vacuum. It's false because, although empty of particles, it's swarming with energy. But what's most interesting about this false vacuum is that even though it's dense with energy, there are no forces of attraction holding it together as exist in a mass-filled space. In fact, there is the opposite: a false vacuum's gravitational force is negative. Everything in it is repulsive instead of attractive. And this false vacuum expands dramatically, doubling in size at the rate of 10^{34} times a second. (That's a 1 followed by 34 zeros, a number so phenomenally large that if you piled up that many pieces of paper it would be 1,000 times as long as the observable universe. So $1/10^{34}$ of a second is mind-bogglingly small.)* There's extraordinary power in doubling a number over and over. An example many people have heard tells of a king who wanted to reward the inventor of the game of chess. The inventor smiled humbly, saying he didn't want much, just a grain of rice on the first square, two grains on the second square, four on the third, eight on the fourth, and so on up to the sixty-fourth square on the board. The king readily agreed, and so was somewhat

*Assuming 100 pieces of paper to 1 millimeter, 100,000,000 pieces to a kilometer, 9.5 trillion kilometers to a light-year, and 10 billion light-years to the observable universe.

startled when by the forty-sixth square or so he required a pile of rice as big as his castle. The sixty-fourth square, it turns out, would have required a pile of rice with some 9 quintillion grains, an amount that would cover the island of Manhattan.[2]

Doubling over and over that quickly let the universe balloon out to sizes we can't possibly conceive of. In some inflationary models, the universe could have expanded from $1/10^{35}$ of a meter to larger than our visible universe in $1/10^{30}$ of a second. Our little observable patch of the universe—all the light that took some tens of billions of years to get to us—would be but a tiny fraction of the whole thing. Guth and his colleague Paul Steinhardt at Princeton have suggested that if the inflation approach is correct, the universe could be 10^{25} (that's 10,000,000,000,000,000,000,000,000) times bigger than what we can actually see.

These very large and very small numbers aside, Guth realized that inflation, as he began to call it, explained away several nagging problems in the big bang theory—namely, why the universe is so smooth (the horizon problem) and why it's flat. The glory of inflation is that no matter how it starts, how dense, how clumpy, the expansion process swells it up into a state that would have to be smooth. Imagine blowing up a balloon, a special balloon shaped like a heart or an animal. As you blow air in, it will go from a very specific deformed shape into a much smoother, rounder shape. But then keep blowing, filling it with air well beyond its capacity. If the walls of the balloon are strong enough, the balloon will keep swelling, losing its expected shape and becoming almost spherical. All the lumps are smoothed out. Dramatic growth, in this case, leads to homogeneity.

Moreover, the sudden expansion naturally leads to a flat universe. Think of the balloon again, pumped full of more and more air, huge amounts of air, until eventually it becomes so large that when you look at its side, you cease to even see the outside curves. It appears practically flat. Inflation took all of space and made it even.

Last but not least, taking a tiny space and swelling it up so dramatically created a universe in which there would be very few magnetic monopoles—perhaps about one per universe. A single monopole may exist out there; we just haven't stumbled across it yet.

Guth hammered out the math of this false vacuum and its dramatic inflation one night in December 1979, staying up until 1:00 A.M. The implications sank in overnight. He knew he was on to something big. The next day he drew a big box in his notebook and wrote in it "Spectacular Realization: This . . . can explain why the universe today is so incredibly flat."

Like any theory, the first version wasn't exactly perfect. In Guth's first inflation theory, different regions of the false vacuum stopped expanding fairly randomly. Each pocket of vacuum would finally grow cold enough to transform from energy into matter, freezing into haphazard chunks. Remnants of this would still be visible today, as various parts of space should be noticeably different as you cross from one such region to another. The fact that we don't currently see these varying bubbles of space was a problem. In 1984 Andrei Linde and Paul Steinhardt adjusted the equations and suggested that our entire universe might exist inside a single one of these bubbles. We don't see these boundaries where the false vacuum froze because the bubbles are so gigantic—much larger than the bit of universe we can see with our telescopes. This version of inflationary theory is one of the main ones investigated today.

Andrei Linde is also known for adding another twist to inflation: a bit of theory called chaotic inflation. This relies on the idea that inflation seems to beget more inflation. As one bubble universe expands, it can spawn new inflationary bubbles, which eventually slow down and develop into an all-new, separate universe. In this scenario our universe was spawned from a previous one and could now be giving birth to new ones constantly; and while each individual universe has a big bang type of beginning, the entire "multiverse" may have had no beginning in time.

Regardless of the problems that needed ironing out in 1979, the inflation theory captured the imagination of scientists fairly quickly. It just answered so many questions, while being based on fairly accepted science of the time.

THE FLATNESS PROBLEM

The inflationary universe offers a good explanation for why the universe is flat, but it doesn't tackle all sides of the issue. First, one needs to understand why the universe is presumed to be flat—after all, we've repeatedly said that scientists have yet to determine the exact shape of the universe. All anyone knows is that it's somewhere close to flat. Maybe exactly flat; maybe a little off. We just don't have the tools yet to measure it exactly.*

To assume a present-day universe that is close to flat means one of several things. Perhaps the universe has always been exactly flat, balanced on this knife edge, the perfect density at which it will never recollapse. Or perhaps it was either greatly denser or lighter in the past. Such a universe would transform over time, changing in density fairy rapidly. It would at some point pass through a phase where the universe was fairly flat—and that would be the time we're in now. We would just happen to be around to measure the universe at the perfect time to perceive the flatness phase.

Scientists are not particularly fond of theories that include statements like "and by some miraculous twist of fate, we measured it at the exact time or place to get a certain result." Copernicus jettisoned the idea that humans were in any special place, and modern scientists don't like to put us at any special time either. So they prefer to believe that the universe has always been pretty close to flat.

Guth's inflationary theory provided a nice reason why the universe might have come out perfectly flat, but there's still a big problem. As scientists map the galaxies, they simply don't *see* a universe that has enough material in it to make it flat. (Remember, this concept about the shape of space is rooted in Einstein's relativity theory, and the universe's

---

*Several experiments may soon change this. Most notably, MAP (Microwave Anisotropy Probe), launched by NASA in 2001 should go a long way toward telling us the shape of the universe.

shape is defined by its gravity, so it's also defined by the amount of mass it contains.) By measuring the amount of light coming from a group of stars, astronomers can make a fairly good estimate of how many stars are there, and therefore how much the entire system weighs. But when you do the math and add it all up, the light adds up to only about 5 percent of the total mass needed for the critical density of a flat universe.

Scientists did not despair. Fritz Zwicky in the 1930s and Vera Rubin in the 1950s had already noticed that something was odd about such mass measurements. The movement of the stars and galaxies as they circled each other simply didn't make sense. As galaxies spun around, the stars on the outskirts moved much more quickly than they should have, given the force of gravity predicted from the amount of matter that could actually be seen through telescopes. Since the laws of gravity are fairly well understood, these kinds of measurements were fairly disturbing. Zwicky, and later Rubin, advanced the idea that there must be some unseen matter, called dark matter, which filled the spaces between the bright stars. We couldn't see this dark material, but we could nevertheless detect its presence.

You might think of being stranded in the woods on a moonless night in the dark on a hill. The fact that you can't see the hill is irrelevant—your body senses the slant of the slope and the feel of gravity's pull. In our own daily lives we trust senses other than our eyes, and so we must do when examining space. (Astronomers discovered Neptune—invisible to the naked eye—in a similar way, since the planet's gravity disrupted Uranus's orbit.)

Most astronomers take the existence of dark matter for granted, but actually hammering down what it *is* hasn't been successful (and no one has ever been able to find as much of it as one would like to see; more on that in chapter 5). There are two basic options for dark matter: something small and fast, possibly basic like a neutrino, possibly exotic and odd, a hypothetical particle we haven't yet detected; or something large and bulky made of nice, normal protons and neutrons, but that for some reason is so dark we can't see it. The exotic particles in the first option have been dubbed WIMPs, for Weakly Interacting Massive Particles, and those in the second category are called MACHOs, for Massive Compact Halo Objects.

Although many people still hold out hope for finding a small-particle dark-matter candidate, nothing has been found so far. One of the more straightforward possibilities would be if neutrinos—small particles that pervade the atmosphere and barely interact at all with anything they move through (billions are traveling harmlessly through your body right now)—turned out to have some mass to them. Since there are so many in the universe, even the tiniest weight would be enough to provide all the dark matter needed for a flat universe. As it happens, scientists studying neutrinos have not proved conclusively neutrinos have mass—and though the occasional experiment says they might, these results haven't been duplicated. We can, however, say that at the very least neutrinos weigh less than one ten-thousandth of an electron. That's good news for us, because if it were much higher than that, they would have contributed so much weight to the universe that it would have recollapsed long before life could have arisen. In fact, if they do weigh that much, they would weigh more than all the stars in the universe.[3] But no one has been able to prove they have weight, and it's fairly unlikely that they do.

Other possibilities for WIMPs include particles that fit in with some of the Grand Unified Theories currently being investigated by particle physicists. Such particles help to create a lovely set of theories describing how all the forces combined into one in the earliest moments of the universe, but like all other theories about those first moments, they are pursued more for their mathematical beauty than for anything concrete that can be observed. These particles are still just a hypothesis, and that leaves us without any WIMP dark matter.

MACHOs, on the other hand, have been found. The basic MACHO categories are large masses such as planets; dead, collapsed stars, known as white dwarfs; or even black holes. In the mid-1990s a group of eighteen astronomers from eight research institutions, led by Charles Alcock, put together a massive search for these dark objects. They observed ten thousand stars a night, watching to see if they changed brightness. If a MACHO passed in front of a star it would be too small to obscure the star; it would, however, cause light from the star to bend around on each side, creating an interesting phenomenon known as microlensing. This lensing makes the star appear brighter because the light rays bend and split around the MACHO, coming to our telescopes from two different directions—appearing to our eyes almost as if they're coming from two different stars.* These "two stars" are nevertheless

*This is not unlike the first proof for relativity found in 1919, when starlight was seen to bend around the sun.

close enough together to superimpose into one brighter-than-normal image.

Some members of the team went into the project thinking they would find nothing and would simply discount the possibility of MACHOs, and then scientists could go back to focusing on the more likely candidates: WIMPs.[4] Instead, over two years they found seven MACHOs, which, considering that they looked at only small portions of the sky at a time, suggested that there were probably huge numbers of MACHOs they hadn't seen. In hindsight this shouldn't have been so astonishing—astronomers have since found a number of Jupiter-size planets circling other stars—but at the time no one had ever detected anything smaller than a star outside our solar system, and the number of white dwarfs surprised just about everyone. In January 1996, at a meeting of the American Astronomy Society in San Antonio, Texas, the team announced that they believed that 50 percent of all the predicted dark matter in our galaxy was made up of MACHOs. And since scientists assume that other galaxies work like ours, the same probably holds true for the rest of the galaxies as well.

So some dark matter has been found—not quite enough to explain why galaxies seem to rotate so quickly, but certainly some. When it comes to the flatness issue, however, there is a slightly larger problem. Even if you ignore trying to actually identify the dark matter, and simply try to guess how much should be there based on the gravity perceived, you don't get enough for a flat universe. The dark matter predicted is only about 20 percent of the critical density to halt the universe's expansion.

All of this is to say that the flatness problem is not quite solved. If inflation theory is correct—and every cosmologist will admit that's a reasonably large "if"—then we have an explanation for why the world *should* be flat, but no proof that it actually *is*. On the other hand, we have the laws of gravity predicting dark matter, which in the simplest models would leave the universe with an omega of 0.2,* not to mention that scientists have been able to identify only about half of *that* dark matter. Astrophysicists nevertheless continue their search for the missing mass, since it seems necessary to solve so many basic confusions about the universe.

THE GALAXY FORMATION PROBLEM

Inflation theory answered one important "why" but then created another question: If everything is so smooth, why do we have galaxies at all? This issue hovered in many cosmologists' minds ever since the first detection of the background radiation. The radiation was so perfectly consistent that it was hard to imagine structure forming. In a swirling mix of evenly distributed, evenly heated particles, why would any of it clump together into planets and stars? That would be like returning to our cup of hot chocolate and finding it co-alesced into chunks of hot chocolate gravel. Even though gravity would naturally pull particles toward

*As we shall see in Chapter 6, there are some modern, more complicated models of the universe in which Ω can be made to be larger than 0.2, and the missing matter is not such a glaring problem.

each other, if the ylem was smoothly distributed, each particle should be pulled consistently in every direction. On the other hand, if there were slight blemishes on this smoothness, they might serve as "seeds" to start pulling more and more particles in. A seed would soon become the size of a basketball, then a building, then a planet. (Alternatively, a huge mass might coalesce and later separate into smaller planets and stars.)

But no one had found any evidence of those seeds. In the years after the first microwave background radiation experiments, the state of the art had moved to flying antennae in balloons. The higher the antenna, the more earth's weather and man-made radiation could be avoided. But these experiments were plagued with problems. Sending an instrument into the air, after all, means you can't fix it if something goes wrong. The instruments had to be prepared and tested on the ground, and then you crossed your fingers and hoped you'd thought of everything that could go wrong before sending it up. A cable dangling in front of the antenna could ruin the whole experiment, and you'd never know until the whole contraption was recovered and brought to the ground. And even if everything did go right, you had only a few hours or so of data—not enough time to truly scan the sky for the tiny ripples in radiation you were looking for.

A team of astronomers at New York's Goddard Space Science Center wanted to solve the problem once and for all. They were finished with balloons. The time had come to put a satellite in space. Orbiting the globe, sending back data to scientists on the ground, this type of instrument would not only avoid random radiation from the earth but also would have all the time needed to collect data from the sky. Led by John Mather, a twenty-eight-year-old who'd just finished graduate school, the team wrote up a detailed proposal to NASA.

The timing was right. As it happened, NASA was looking for this kind of project, one that would tackle solid, basic science but that could be accomplished far better in space than from the ground. Add that the origin of the universe would capture the imagination and interest of the public and this was a perfect experiment. Two other astronomy teams had made similar proposals, and NASA pulled team leaders from all of them for one big effort: a satellite named COBE, or Cosmic Background Explorer.

COBE (pronounced "cobey") was to have three main instruments: one to measure the background radiation a hundred times better than had been done previously; one to search for the faint glow of the first galaxies; and one to seek out those tiny ripples, the first seeds of gravitation that everyone hoped would be there.[5] Fifteen years and $60 million worth of engineering and testing later, COBE was finally launched on November 18, 1989.

A mere two months later, the team announced their first results. The background radiation was there—perfectly distributed in every direction, with exactly the spectrum one would expect from the big bang theory, hovering at a temperature 2.726 degrees above absolute zero. John Mather stood in front of a packed room at an American Astronomy Society meeting in 1990—this one in Crystal City, Virginia—and showed a graph so perfectly aligned to what the big bang theory predicted that he received a standing ovation. The COBE data were definitive. It was the proof. It is the data that every adherent to the big bang cites as his or her reason to believe.

But that smoothness still had people worried. The COBE team was narrowing the antenna's focus, measuring the temperature over smaller and smaller ranges, trying to find some—any—discrepancy. By April 1992 the team was ready to make their second announcement: they had found temperature fluctuations. The difference was only thirty millionths of a degree hotter than the surrounding background radiation, but these hot bits were scattered around the sky, some the size of a constellation, some larger, one even covering a fourth of the sky.[6]

The press had been alerted, rumors spread among the scientists, and the announcement—this time at an American Physical Society Meeting in Washington, D.C.—was again made to a packed hall. Showing a colorful image of the hot spots in the heavens, one of the scientists on the podium, George Smoot from Berkeley said, "If you're religious, it's like seeing the face of God." Later, Stephen Hawking said it was the greatest discovery of the century, if not of all time.

The excitement was infectious, and looking back the hype may have been more than the announcement deserved. Certainly there are many scientists who groan every time they're reminded of Smoot's overzealous remark—invoking the divine just adds insult to injury. It probably wasn't the greatest discovery of all time, but it was a major find, especially since without evidence for temperature fluctuations the big bang theory would have come under heavy attack. (In fact, many people have pointed out that COBE really would have been one of the most important experiments of the century if it *hadn't* found the background radiation or the temperature wrinkles. That would have turned cosmology on its head.)

Finding those wrinkles in the sky was crucial for the survival of the big bang theory. The wrinkles deserved much of the press they got. But the irony is that they didn't really solve the question of how galaxies formed. Those swaths across the sky—they are so big, so much larger than a galaxy—that theorists are hard pressed to figure out how they might have seeded stars and planets.

The best theories of the creation of structure in the universe have to call upon dark matter again. There must have been some dark matter to create enough gravity to pull particles together. Scientists have investigated two options: hot dark matter ("hot" because the particles moved so very fast, and temperature is defined by the speed of particles) and cold dark matter. The hot dark matter would likely be something like a neutrino with mass, while cold dark matter would be some other kind of exotic WIMP. By hypothesizing versions of the early universe that match the COBE data, and then adding the required exotic dark particles, cosmologists could go to their supercomputers and see how their cyberuniverse would form.

All hot doesn't work. Hot dark matter universes evolve on computers into large galaxy cluster sizes first, then into smaller forms such as galaxies, and finally stars. Since the evidence suggests that the process works the other way—stars form and then slowly group together into larger clusters—the all-hot-dark-matter scenario was discarded.

But all cold doesn't work either. Starting from an early universe structured the way COBE presents it, the cold-dark-matter universe won't evolve into reality as we see it through our telescopes today. And if you get a cold-dark-matter universe to appear like ours, which is possible, it never passes through a phase that matches the COBE data.

In the end, the best scenario seems to be a mixture of the two, despite the fact that no one has yet detected either kind of dark matter. And to top it off, this mixing and matching doesn't sit well with most theorists, who continue to search for a simpler explanation.

A couple of other ideas do get floated around from time to time about how galaxy formation might have occurred without the benefit of seeds of dark matter. If one doesn't accept inflation theory, for example, then one can assume an irregular early universe instead of the more common hypersmooth one. In such a clumpy universe, any particularly dense area would collapse into a black hole. Eventually the whole universe would be filled with black holes—whatever wasn't near enough to collapse into their gravity would nevertheless distribute itself in galaxies of matter around the black hole. These clumps would in turn seed galaxy formation.

Another possibility is that phenomenally large heated gas clouds could have endured explosions that compressed matter into all the stellar material we see today. Since astronomers believe that stars form today out of such explosive clouds, it's not such a stretch to imagine an extroadinarily large one exploding into a whole galaxy.[7]

WHAT'S REALLY BEEN SOLVED?

Historically, there has been a certain glossing over of many of the questions the big bang theory has brought up. Inflation solves certain problems but is devoid of proof even as solid as that for the big bang. Dark matter offers hope that the universe is at a critical density, but it's mostly math and gravity—no one has completely hammered out the particulars yet. And COBE's temperature fluxes are so large that they don't really correspond to galaxies per se.

Many astrophysicists nevertheless back the basics of much of these theories—while admitting that the details need to be worked out—almost as heartily as they do the big bang. Inflation theory is, in a word, beautiful. In one simple stroke it offers an explanation for why physics as we know it had no choice but to create a universe like the one we live in. There is no fudging, no ad hoc assumptions to throw in—it's a theory that makes sense.

But there is no denying that this "beauty" (or simplicity) criterion is a subjective one to hang one's theories on. Critics of the big bang theory scream loudly that there are so many holes and so many unexamined assumptions in these theories that they should all be viewed with a lot more skepticism than they currently are. History has certainly shown that many brilliant thinkers can be led astray by taking certain "facts" for granted simply because they appeared to be "obvious" truths.

On the other hand, supporters of the big bang respond that past theories—Aristotle's circles, Newton's perfectly balanced cosmos—didn't have the benefit of rigorous science and advanced telescopes. Modern assumptions, they insist, are not arbitrary, but based on state-of-the-art experiments.

And what scientists don't know for sure yet, should soon be known. Major experiments are under way to measure the background radiation a thousand times better than COBE did. We should soon know the exact value of omega and the temperature fluctuations.

As these new data come in, no one is going to be unduly surprised if the basics of these concepts—inflation, missing matter, galaxy formation—need to be substantially remolded or scrapped altogether.

On the flip side, should some proof, some bona fide evidence come in, these theories will be accepted as wholeheartedly, and as quickly, as the big bang. In much the same way, it will not be the clinching experiments that give the proof its robustness, but the way the theory makes sense in the minds of those who hear it.

5

Glitches

There's no doubt that the big bang theory is accepted so universally because it is taught essentially as fact. We've all learned the earth is round, and few ever think to try to prove that detail for ourselves. When the same science teacher tells you the universe began with a bang, most people accept it as readily. The majority of the world population that accepts the big bang theory does so unquestioningly.

But just because it's dogmatic doesn't inherently mean it's wrong. It has been said that cosmology in any era is dominated by a few loud voices. The echoes of Aristotle's booming voice lasted centuries, and even now the names of those who hammered out the big bang theory are linked to large personalities. But personalities aren't enough to make or break a theory. They're crucial to disseminating a hypothesis, making it popular, getting the word out. Science is definitely more subject to personality quirks than most scientists would like to admit, but it's also clear that science is more than that. Unlike in the 1950s and '60s, when picking your cosmology of choice was a matter solely of preference, the decision can no longer be a "gut" choice.

The theories *do* need to jibe with the facts as they are observed. Take Fred Hoyle, gregarious supporter of the steady state theory. No matter how loud his voice, it couldn't drown out the fundamental fact that his theory could not explain background radiation. Scientific theories are based on more than unquestioned faith. The big bang rests on some valid observations, observations that only this theory—of all the scientific theories about the universe that humans have ever devised—can explain. Before jumping into the problems with the big bang theory, it is first important to understand where it succeeds.

THE SUPPORT

There are basically three supporting facts: the amounts of deuterium and helium in the universe today correlate to the amount the big bang would have created in the first minute; we've detected a wonderfully smooth background radiation that corresponds to the temperature we would expect to have been produced three hundred thousand years after the big bang; and the universe is expanding.

The helium itself isn't as important as the deuterium. One could change the standard big bang theory quite a bit and still produce the amount of helium we see today; but deuterium is finely tuned to the initial conditions of that explosion. Today deuterium is created in the hearts of stars, but it is instantly destroyed, so none escapes. No other process has ever been discovered that makes long-lived deuterium, so it's believed that every atom of deuterium in the universe today was created in the first minute after the big bang. (Fred Hoyle, you may remember, hypothesized that the light elements may have been created inside massive black hole–like objects, but most scientists believe that relying on these as the furnaces to cook up deuterium doesn't make sense.)

The background radiation is equally hard to explain by any other theory, although not entirely impossible, and it's certainly been tried. One option is that energy from the earliest stars (formed in a non–big bang universe) was absorbed uniformly by particles everywhere, only to be sent forth again as the even background radiation.[1] But it's not easy to make a mathematical model of this come out with the incredible smoothness we detect—smoothness that the big bang version predicts as a matter of course.

The last piece of supporting evidence for the big bang theory is, of course, that the universe is expanding. But while that fits in with the theory, it certainly could be explained by other models. The steady state theory, after all, incorporated it perfectly well. Expansion fits—indeed, is necessary—but doesn't do much to elevate the big bang theory over and above any other model.

These are the firm bricks providing the foundation for the big bang theory, bricks that justify accepting the big bang as the best model ever found for how the universe came to be.

But as soon as one starts delving into the details—just what kinds of material were created in the first explosion, how that led to the formation of stars and galaxies, why the entire universe didn't disappear within minutes after it started—some glitches start to appear.

THINGS NOT UNDERSTOOD

One issue that shows up in most press reports about the big bang is that the theory posits a fairly young age for the universe, in some versions as low as eight billion years. Compare that to certain stars that are thought to be over fifteen billion years old and you appear to have a serious problem. However, this never concerns astronomers all that much. Over the 1990s, the numbers converged more and more to around 13 billion, and while some different values still come in, the numbers bandied about have large enough uncertainties in them that it's assumed the problem isn't too great. It's a bit of a mystery, but no one is going to pull the whole theory down based on such a small discrepancy.

That said, the denser the universe is, the younger it has to be. So requiring a fairly dense universe, one where there's enough dark matter for omega to equal 1, as so many theorists like to do, makes this age discrepancy worse. A high density keeps the age of the universe at the low end of the spectrum. It's hard to reconcile that precious critical density with stars of the right age.

But these are just the beginning of the issues with dark matter. The fact is that the dark matter problem is reaching something of a crisis, although few astronomers have been willing to admit this yet. Forget not finding any ideal dark matter candidates. The problem isn't that no one can find the missing matter (although they can't) but that even if theorists stomp their feet and shake their heads, observations haven't even shown that the universe is at the critical density.

Recall that the amount of material we can see in the universe amounts to only about 5 percent of the critical density. The movement of stars and galaxies, twirling at speeds faster than you'd expect based on the amount of mass we can see, makes a good case that there is something dark and heavy lurking in the heavens, adding more matter. This is fairly well accepted. The problem is that once you do the math, even if you interpret the data extremely liberally, trying to boost the amount of dark matter as high as possible, you still reach only the 20 percent mark. We don't detect anywhere near enough mass in the universe to match the inflationary theory requirements.[2]

Astronomers examine the shape of the universe with other techniques as well. Not content to simply measure the density of the universe based on what they can see, they use other tools to measure the very geometry of space and how fast the universe expands. One basic trick is to look at galactic clusters at incredible distances to get a sense of their age. If hugely distant objects (implying they traveled a fairly long distance) are similar in age to our own galaxy (implying they've been traveling for a fairly short time), then one can safely assume that the universe has expanded very quickly. Alternatively, if they're a great deal younger (traveled a long time) and closer (traveled a shorter distance), you assume that the universe is expanding slowly. The first scenario implies a low-density universe (since high density slows expansion), and that is what has been found by a variety of researchers using a variety of telescopes. Evidence from these kinds of tests has piled up. They see a consistently low-density universe with about 20 percent of the mass needed to bring the expanding universe to a slow, gentle halt.

More than anything, this kind of data affects the inflationary theory—an add-on to the big bang theory; a hanger that, if disproven, doesn't require dismantling the entire standard big bang model. But the dark matter problem affects the basics of the big bang model, too. Without enough dark matter there's no way for galaxies to have formed in as short a time as they seem to have done. Dropping dark matter out of their models would make it impossible for theorists to understand how a universe could get from the big bang to what it looks like today.

Ditching dark matter has other problems. Abandon a flat universe, abandon the inflationary theory, and the horizon and flatness problems rear their heads. We'd also wonder where all the monopoles were, and we wouldn't have any explanation for the first moments of the universe. Without a workable theory to replace it, most scientists are content to work with what they have, modifying the inflationary theory until they have reason to believe it well and truly dead. (Or proved!)

Dark matter is crucial for galaxy formation, but galaxy formation has its problems even without the issue of the missing matter. While COBE showed variations in the heat background, thus showing that the early universe was not perfectly smooth and did indeed allow for seeds of galaxy formation, those ripples were nevertheless huge. They're so large that scientists can't figure out how they turned into the kind of structure we see today. This doesn't mean that there may not be smaller ripples, simply that COBE couldn't see at the level of detail necessary. Enter a NASA project called MAP (Microwave Anisotropy Probe), launched June 30, 2001. MAP's job is to follow up on COBE's work and provide the best images of the microwave radiation to date.

Other projects to map the radiation—and there have been a number, relying on flying balloons instead of launching an instrument into orbit—have come up with other complications. Some show an "echo" in the radiation—an extra peak that would correspond to the vibrations as the radiation bounced around the universe, much the way a sound wave traveling through an organ pipe will resonate at a variety of frequencies. This is a comforting peak, one that hits just where math predicts it should. But other instruments haven't found the peak at all.

MAP, which will collect data for two years (adding up to as much data as COBE could have collected in four thousand years), is set up to hammer down this peak or declare it missing once and for all. MAP will be able to see ripples in the radiation far smaller than COBE could find. Finding these small fluctuations is fairly crucial—it's these fluctuations that would act as tiny gravitational seeds to start the process of a whirling gas turning into a star, leading to many stars, galaxies, and clusters. The hope is that with a map that precise, a "universe" on a supercomputer starting from those parameters truly would evolve into a model that corresponds to the reality we currently see.

Of course, far from solving these problems, MAP might just add more complications. It might not find that echo peak, or might not see smaller fluctuations, or might show something entirely unforeseen and unexpected. (In many ways that would be a lot more interesting result. If something fundamentally new shows up, it will give theorists a starting point to rework the glitches in cosmology.) Whatever questions MAP raises will likely be tackled by an instrument called PLANCK being assembled by the European Space Agency and due to be launched in 2007.

ALTERNATE INTERPRETATIONS

While everyone agrees that there are a few unanswered details in cosmology, some take it even further. Some claim that the whole big bang theory is wrong. Many of these people strike at the very heart of the big bang theory: expansion. While, as mentioned earlier, an expanding universe doesn't require that the universe began with a bang, the big bang theory certainly requires an expanding universe. If it turns out that galaxies and stars aren't receding from each other, then the entire theory would fall apart.

One of the most well-known attacks on expansion is the tired light theory. Remember, the best proof of expansion is that the farther away an object is, the faster it seems to be moving. Scientists determine that speed by examining the color of the light—the more redshifted it is, the more of a Doppler effect we believe we're seeing, and the faster it's thought to be traveling. But this requires a few basic assumptions: at the most fundamental we assume that the laws of physics seen on earth are the same laws that apply to the entire universe. Scientists study objects billions of light-years away and assume they work the same way they'd work on earth.

But what if light itself, as it traveled such vast distances through space, lost some energy? Losing energy, the light would appear redder (since less energy implies a longer wavelength, which implies redder light) even if it came from a star that was standing still. The whole basis for believing the universe was expanding would disappear.

The rebuttal to this rests on the fact that scientists know a fair amount about light. The speed of light is an integral part of relativity theory, and as fast as it travels, it's nevertheless been measured exactly to an accuracy of one part in ten million. Since so much in the universe does work exactly as we think it should based on relativity—such as the way light bends around large stars, or the way certain stars shine—most scientists don't think we should scrap our fundamental beliefs about the way light behaves.

Another attack on redshift theories comes from Halton Arp at the Max Planck Institute for Astrophysics in Germany. Arp is sort of a champion for irregular galaxies. In the 1960s he put together the *Atlas of Peculiar Galaxies,* thinking that studying these would help in understanding galaxy formation. When quasars were discovered, he noticed that many of them congregated around his odd galaxies. Of course, when one looks into the sky, even with the strongest telescopes, one can't tell how far away an object is. Astronomers rely on a variety of tricks to determine distance, such as the variable Cepheid stars that Hubble used to measure the distance to other galaxies, or by measuring redshift. Any of these quasars might be millions of light-years closer or farther away than these galaxies, but the fact that several consistently congregate right in the line of sight with these peculiar galaxies suggested to Arp that they were connected. And yet many of these quasars had vastly different redshifts from each other, which in the standard interpretation would suggest that they were at very different distances from the earth. On one hand, Arp's data suggested that these quasars were about the same distance away; on the other hand, the redshifts made it appear as if they weren't. Most scientists believe the redshifts; these quasars, they think, are nowhere near each other, and are moving at different rates, so it's perfectly reasonable for them to have different redshifts. But Arp believes in his galaxies. If all these objects are at the same distance, then it seems clear the redshifts themselves aren't as good a tool for determining how far away an object is as previously thought. Arp believes that stars and galaxies have an intrinsic redshift, an intrinsic color that has nothing to do with whether the stars are moving away from us.

These Arpian objects, as they're called, have never undergone much rigorous study by anyone else and certainly haven't been accepted by much of the scientific community. (Arp has strongly argued, perhaps correctly, that the scientific community hasn't accepted the data because the mainstream scientists are so resistant to contrary opinions that it's nearly impossible to get contradictory papers published.) Nevertheless, it's important to understand the implications if Arp is correct: it is believed that the redshift corresponds to how fast an object moves away from us. From this information Hubble described a universe in which objects farther away were moving faster, and therefore we believe the universe is expanding. Astronomers accept this so completely that they have begun associating high redshifts with objects that are very far away and very young. Redshifts are not, in and of themselves, a sign of a star's age or distance, and yet redshifts have become intrinsically entwined with how we determine not just the speed of any given object, but also how old and how far away it is.

If the interpretation of redshift is wrong, then all the proof that the universe is expanding will disappear. It would undermine everything that's been mapped out about the heavens. Not only would the big bang theory come crashing down, but scientists wouldn't be able to determine how the nearest galaxy is moving, much less how the whole universe behaves.

ACCELERATION

For the moment, however, expansion seems to be fairly safe, though it did get a jolt in 1998. Since Hubble, it has always been assumed that the universe is expanding but slowing down. This makes sense when you think about an explosion: The initial burst of energy gives everything speed, but as all that material travels, the gravitation of each neighboring object helps slow it down. While gravity might never be enough to slow it down completely, it's nevertheless going to slow.

In 1998, however, two independent observations mapping distant supernovas showed that the universe might be accelerating outward, the very vacuum itself causing ever faster expansion, the way that first false vacuum is thought to have done in inflation theory.

The only way to explain this kind of acceleration is to put the cosmological constant back into Einstein's equations. (That's the bit Einstein once called his "biggest blunder," originally put in to maintain a static universe. Technically it's never been taken out—the cosmological constant is always part of the equations; it's simply assumed to be zero.) This rubs many theorists the wrong way. Incorporating an almost arbitrary number into otherwise simple equations—and remember how much scientists like to keep their explanations simple—just doesn't sit well. Nevertheless, many cosmologists are now attempting to rework what they know about the big bang theory to incorporate this mysterious extra quantity. (More in chapter 6.)

It's fascinating to note, though, that when this incredible fact came through—a fact that turned everything we thought we knew about the universe upside down—there was no consensus to chuck the whole big bang hypothesis. The first step was to figure out how to accommodate this new fact about an accelerating universe into the established theory. It goes to show how solid is the mainstream belief in the big bang.

More than that, over the next few years this fascinating acceleration became incorporated in such a way as to help support the cosmology theories in existence. For one thing, the cosmological constant is explained as energy in the very vacuum of space itself. Though technically "nothing," the vacuum has an antigravitational repulsion, giving it an enormous burst of energy to let it accelerate. This is, of course, the same repulsive force thought to have driven inflation in the first moments of the universe. If we can watch this force in action in today's universe, it gives just that much more credence to the possibility that inflation was driven by the same mechanism. This new manifestation of anti-gravity has forced staunch skeptics of the inflationary theory to admit that it's a little more likely.

Even more importantly, this cosmological constant is now seen as perhaps saving cosmology from the missing dark matter problem. Theorists, of course, like the idea of a universe that is exactly flat—the kind of universe in which the shortest distance between two objects is a straight line and parallel lines never cross. (The earth, for example, isn't like this at all, which one can tell from spending a few seconds with a globe—all those "parallel" longitude lines converge at the poles). In the simplest model of a flat universe—the model we've mentioned so far in this book—the "flatness" comes from having a very precise amount of matter. We see only about 5 percent of that amount in the stars and galaxies we observe. We see another 15 to 25 percent of dark matter, which we can infer based on how the galaxies and stars move. But that still leaves a whole bunch of matter unaccounted for. Einstein's relativity equations do not, however, require that matter be the only thing that creates flatness; the cosmological constant can help. So the acceleration of the vacuum might actually save cosmology theories by being the missing piece that creates the flat universe that theorists crave. The cosmological constant might add the 70 to 80 percent needed to bring omega up to 1, thus saving inflationary theory and most of the rest of cosmological theories.

That's the good news. The bad news comes from one of those ubiquitous "whys." Why would the cosmological constant be just that size, just the perfect amount to create a flat universe? The accelerating universe might solve one problem, but it opens up another series of questions for theorists to try to solve. They would have to set about understanding the physics laws that guaranteed such a coincidental occurrence.

OTHER OPTIONS

Rallying against the big bang theory until he died was steady state theorist Fred Hoyle. Along with Geoffrey Burbridge and Jayant Narlikar, Hoyle proposed what they call the quasi–steady state theory, a theory they began crafting in 1993. The theory continues to rely on minicreation events, a constant invention of particles that occur, on a smaller scale, much the way inflation is thought to have created the whole universe at once: a combination of negative energy and positive mass flings newly born particles out into the cosmos. Quasi–steady state claims that these jets of just-created matter are what build up a black hole, so a black hole is built from the inside out instead of, as is conventionally believed, solely by the hole's massive gravity scooping up all that comes too close. It is only in this way, believe the three scientists, that black holes could grow massive enough to account for how fast galaxies twirl around them. To explain away the standard supports for the big bang, the theory claims that the all-important cosmic microwave background radiation is merely diffuse starlight that has equalized around the universe over time and that all the light elements could have been produced in the fiery hearts of stars.[3]

Various other dramatically different theories come up from time to time, and they're interesting enough to mention, but there's little to judge any of them by. There's no way to tell if they're accurate. For example, Hoyle and Narlikar pointed out that we would perceive the universe to be expanding even if space in fact stays the same, but matter shrinks.[4] In this topsy-turvy world, we'd still perceive galaxies to be getting farther and farther away, but—conveniently for Hoyle, who abhorred the single moment of creation—we do away with the initial singularity. All matter could have been here eternally. But there's no real way to test this idea, so it remains just an intellectual exercise.

Another alternative to the big bang is the plasma universe theory proposed by Hannes Alfven, a Nobel Prize–winning astrophysicist from Sweden. Strong opponents of the big bang theory often jump to the plasma model as their favorite alternative. Plasma is the fourth state of matter, after liquid, solid, and gas. It's an electric soup heated high enough for electrons to get stripped off the atoms, so that the gas glows. The shimmering light in a neon sign is a plasma, as are the Northern Lights. In the plasma universe theory, there is no definite point in time when the universe was created; it has been around forever, and it started out as a uniform plasma. After trillions—*trillions*—of years, tangled electromagnetic plasma strands began to seed galaxy formation. The plasma universe explains the background radiation as its own energy reemitted from interstellar dust, and expansion as the repulsive force when groups of antimatter and matter collide. Unfortunately, experiments to differentiate between this theory and the big bang theory are hard to come by at the moment. One strike against the plasma theory, though, is that for us to see all other galaxies receding so evenly from us, we must be at the very center of a matter (or antimatter) region of the universe. Scientists tend to shy away from any theory that requires us be in a "special" place in the universe, and so many dismiss the plasma theory on this basis.

Another alternative to the big bang is that instead of our universe being created at a single moment in time fifteen billion years ago, it has always been around—going through a series of bangs, expanding, and then crunching all the way back to a point again, only to reexplode into the cosmos. There's something compelling about this image—it's a way around there having been a creation out of nothing—but it doesn't work particularly well scientifically. For one thing, no one has been able to make the laws of physics conform to this idea. No known process could make a crunch reverse into a bang. In addition, we'd expect to see extra heavy elements in the universe left over from the last cycle. Since the big bang works well to describe the amount of elements we see, few cosmologists deem it necessary to monkey with it.

One problem with these alternatives to the big bang theory is that few can account successfully for the cosmic background radiation. This radiation in and of itself doesn't require that the big bang theory per se be correct; however, it's hard to explain away the radiation with anything other than some historic fireball—that radiation seems explainable only by the kind of even heat described by the big bang. A couple of alternate explanations have, of course, been made: Burbridge, Hoyle, and Narlikar, as mentioned, suggest that starlight itself, as it traveled around the universe, constantly being reabsorbed and reemitted by various objects in space, could over time thermalize into a smooth background radiation. Most physicists, however, would insist that whatever modifications the big bang theory undergoes, the existence of the microwave background radiation ensures that it will always incorporate some primeval fireball.

CONCLUSIONS

What it comes down to is this: Our standard cosmology theory does not explain everything perfectly, but no one can disprove it, and no other theory explains the facts any better. Some alternate theories may explain the facts just as well, with about the same number of glitches as the big bang theory, but a theory needs to be clearly *better* to unseat the current paradigm. Throughout history, society has never overturned a paradigm without having another one to replace it. The big bang is no different—it will survive unless something fundamentally better is derived, and that has yet to happen.

Moreover, if you leave off the add-ons to the theory, such as inflation and galaxy formation, the big bang itself—the part where the universe is expanding from a much smaller entity, and there was a gigantic fireball at some point in our history—seems fairly well proven. So while cosmologists admit they haven't solved everything, most would say that even if a new theory comes along, it will include many of the big bang and inflation ideas currently in vogue. In the same way that Einstein's relativity incorporated Newtonian gravity—Newton's gravity wasn't *wrong,* just incomplete—new theories will not deny the previous theories, simply improve on them.

6

Current and Future Research

The scientific method, of course, has a prescribed way of dealing with glitches in a theory through experiments and observations. Cosmology theories, however, began at a disadvantage—even twenty years ago, the tools to tackle many of the problems of the big bang theory couldn't even have been envisioned. As such, cosmology has often been filled with fervent beliefs but not that many facts.

But technology improved dramatically by the end of the twentieth century. Couple that with determined physicists who insisted that cosmology should be brought from the world of the speculative into the world of science, and the field is full of new, almost futuristic experiments. The strategy to knead out the kinks in cosmology is well on its way.

MAKING THINGS COMPLICATED

One basic issue that has dogged cosmology is that while the big bang theory is fairly simple, the universe is complicated. Despite astrophysicists' best efforts to insist on simple theories—theories with only one kind of dark matter, or theories with a universe at a perfectly critical density—the cosmos isn't willing to comply. For example, as mentioned before, cosmologists believe for a variety of theoretical reasons that the universe is "flat" (also known as a universe where $\Omega = 1$), and this should correspond to a very specific amount of mass in the universe. But scientists haven't been able to find enough mass to make that model work. This problem was nearing a crisis between theory and experiment until some new data managed to complicate the issue, and by discarding the simple model they'd been working with, cosmologists may have recently solved the problem of the missing mass.

The original model went approximately like this: If the universe is "flat," meaning that parallel lines always remain parallel, and the fabric of space doesn't curve the way a sphere or a saddle does, then a certain constant, Ω, will be 1. At its most basic, Ω only describes this shape, but it's entwined with other concepts too. A value of 1 for Ω also was associated with the universe's being at some critical density; that is, it had the exact amount of matter that would cause the perfect amount of gravity so that the universe would forever slow its expansion, coming to a gentle halt but never recollapsing. So Ω has been used to represent both the curvature of space and its density.

But this works only for the simplest models. Albert Einstein, you may remember, originally introduced a "cosmological constant," a repulsive force that counteracted the force of gravity, keeping the universe from collapsing, to keep it stable. However, Aleksander Friedmann, the Russian meteorologist who studied Einstein's relativity equations enough to suggest several different possible universes, soon pointed out that no universe governed by relativity could ever be in stasis. It wasn't going to stay still even if this mysterious repulsive force existed. In fact, Friedmann's models showed that when you included the cosmological constant, all sorts of complicated universes might arise. And in these models, a universe that was flat, with $\Omega = 1$, might actually contract or accelerate instead of living out a predictable life of simply slowing down at a nice, steady rate over time.

Let's explain that a little more. The shape of the universe, as previously mentioned, is defined simply by how one moves through it. Two people moving in parallel paths in a straight universe will never bump into each other. Those same two people traveling along seemingly parallel longitude lines on, say, the earth, will eventually converge at the north pole, the difference being that these two people are walking on a curved surface, no matter how "flat" it appears to them. But relativity says that the curvature of space depends on the amount of mass, and therefore gravity, it contains, so density will naturally contribute to the size of Ω. But if the universe has some subcritical amount of mass, as ours seems to, which all alone would make the universe saddle-shaped (Ω less than 1), it might nevertheless be flat if there is a large enough cosmological constant. If the universe is not just expanding but accelerating with some mysterious force of negative energy—much like the negative energy that fueled inflation—then the universe could have very little mass and yet be flat, yet have $\Omega = 1$.*

Confronted with the persistent lack of enough dark matter, certain farsighted theorists suggested that invoking the cosmological constant might be the best way out of the dilemma. And early in 1998, a handful of seven-billion-year-old supernovas showed that they just might be right.

A supernova is the brilliant explosion that occurs at the end of a star's life. Tycho Brahe, remember, spotted one in 1572, and thus learned that the heavens were mutable. Two other supernovas were seen by historical astronomers, one documented by both Chinese and Native Americans in 1054, and the other by Kepler in 1604. Not another supernova was seen until this century—not because they weren't there, but because they are fairly rare and one never knows where to look. A typical galaxy experiences a supernova about once every three hundred years, so one would have to watch some five thousand galaxies constantly to spot a couple of supernovas a month. It is only with modern telescopes and photography systematically cataloging sweeps of the sky that astronomers have been able to find enough supernovas to learn anything interesting about them.

---

*A little math: Historically everyone has thought of Ω as simply the part of Ω that comes from the density of the universe. Now that people seriously consider the possibility of a cosmological constant, scientists distinguish between Ω_M, the contribution from matter, and Ω_λ, the portion contributed from the cosmological constant. Therefore the Ω mentioned here, the Ω that when equal to 1 means the universe is flat, is now the *total* Ω, the sum of Ω_M and Ω_λ.

The most important interesting thing they've learned is that certain types of supernovas—dubbed Type Ia supernovas—are what's known as standard candles. Like the Cepheid stars that Edwin Hubble used to measure the distances to nearby galaxies, a standard candle is any object the intrinsic brightness of which we can measure from earth. By comparing its intrinsic brightness to how bright it appears to us on earth, we can measure how far away the object is. Type Ia supernovas are the brightest of all supernovas, taking about three weeks to reach the peak of luminosity and then fading slowly away for months. The time it takes to peak and fade is related to how bright the supernova is at its brightest, so by watching the supernova change over time, astronomers can measure its intrinsic brightness and consequently its distance.

Two groups of scientists in the mid-1990s—one known as the Supernova Cosmology Project, the other as the High-z Supernova Team—examined supernovas that were four billion to seven billion light-years away. At such distances these supernovas would also be fairly seriously redshifted, meaning that they were moving quickly away from us in accordance with Hubble's law that the universe is expanding and more distant objects are receding faster than closer ones. Both teams found, however, that when they compared the amount of redshift to the distance, the supernovas were significantly farther away than the redshift indicated. In other words, the supernova was moving, but not as quickly as it needed to be to get as far away as it is. Of course, looking at a supernova that is seven billion light-years away means we are seeing the supernova as it looked seven billion years ago. So that redshift, or really that speed of the supernova, is the speed it was traveling when the universe was half the age it currently is. And yet it's much farther away from us than it should be if it has been traveling at that speed consistently throughout time. It would be like watching your sister drive a car out of your driveway at ten miles an hour and getting a phone call an hour later that she was now in a town sixty miles away. You'd know there was no way she could have driven at ten miles an hour the whole time. The only way she could have gotten there so fast was if she had sped up somewhere along the way. In the same way, these supernovas must have sped up somewhere along the way.

But that kind of speeding up has never been part of simple cosmological models. The universe has been thought to be expanding—but to be slowing down. The supernova data suggested that the expansion of the universe was, in fact, getting faster over time. The universe is accelerating.

And while the initial shock to many cosmologists was great, they quickly saw that perhaps this acceleration was the answer to their dark matter problems. By putting in some repulsive force that overcame gravity and let the universe expand dramatically, the universe could be geometrically flat, even without the critical amount of mass. Ω could be made to equal 1, even without enough dark matter.

Dark Energy

Thinking they had a solution to their problems was the scientists' first response. But, of course, the details had yet to be worked out—and they're still being adjusted. For one thing, when the math was done to find what the cosmological constant should be via theory, it was 10^{120} (that's a 1 followed by 120 zeros) times bigger than what we actually witness. A cosmological constant that large would mean that everything in the universe should be expanding so quickly that you would not be able to see beyond the end of your nose. So right off the bat, theory still isn't jibing perfectly with observation. It's clear that how gravity and antigravity work isn't understood well at all.

Next, a true "constant" didn't particularly work either. If this repulsive force were strong in the early universe, then it would have driven all the particles away from each other, and there's no way stars and galaxies could have formed. So this force that is sending the universe reeling must have changed strengths over time. Since this clearly isn't the cosmological constant as originally envisioned, a new term was coined for this mysterious dark energy: quintessence. The term comes from the ancient Greeks' fifth element—after air, earth, fire, and water—a heavenly element that was supposed to be perfect.

What makes up the quintessence isn't completely understood yet either. In inflation theory, the repulsive force stems from the vacuum itself, but if that were the case here, then the force would have been too strong in the past. There's always been a lot of empty space around, even in the past, and the vacuum would have caused a repulsive force that was much too strong—again there would be the problem that particles wouldn't have been able to coalesce into stars. So what causes this repulsive energy is still unclear.

Nevertheless, by invoking quintessence, physicists can theorize a flat universe that corresponds to the observations astronomers make with their telescopes. The amount of matter in the universe might add up to only 30 percent of the critical density, but add the quintessence and you can still get $\Omega = 1$, you can still get a flat universe. That would mean the bulk of energy in our entire universe comes from this bizarre, heretofore unknown force.

Like many adjustments to the basic hot big bang model, the existence of quintessence raises some of its own questions. Since the strength of quintessence changes over time, the universe used to be dominated by matter, simply expanding slowly, and only recently began to accelerate. The question, then, is why it has only just started accelerating. As with so many cosmological concerns, scientists are wary of suggesting we are in a special place in time or space. Why should we be experiencing this acceleration at this specific moment?

One solution, advanced by Paul Steinhardt, is that a matter-dominated universe for some reason automatically triggers the quintessence at a certain point in time. The mass in the universe slows down its expansion long enough for gravity to do its work, to allow large structures to form. Something in that state then leads automatically to the repulsive force of quintessence, which grows in strength until the universe becomes dominated by this dark energy instead of matter. So any universe that starts as ours did—that has the laws of physics ours does—will automatically go from decelerating to accelerating at just about the time when stars, planets, and galaxies have been fully formed. There may even be a future trigger, when the strength of the quintessence reaches a set amount and reverses the process, bringing the universe back to one that is matter-dominated.

While cosmologists like Steinhardt are busy creating models of the new quintessence, many are also devising ways to study the phenomenon more closely. A group of scientists have proposed a new satellite, dubbed SNAP for Super-Nova/Acceleration Probe, that could study more supernovas than have ever been tracked before—on the order of two thousand a year. SNAP hasn't been funded yet, but it has a major groundswell of support from numerous cosmologists who believe it will help determine just what causes this mysterious dark energy.

MAP

While SNAP is still merely a gleam in scientists' eyes, some fairly large experiments will soon turn in amazing amounts of new data. Launched in the summer 2001 and with results expected in fall 2002, MAP, which stands for Microwave Anisotropy Probe, is the heir to COBE. The MAP satellite will measure the cosmic background radiation with thirty-five times more detail than COBE did, and that level of specificity will give volumes of information on just about all the major cosmology questions still unanswered.

At the simplest, we will be able to see density fluctuations that are much, much smaller than the gigantic spots seen by COBE. It's believed that these small ripples are what helped seed stars and galaxies. If they can be mapped out in detail, then we should have a much better understanding of how these large structures formed.

Mapping the background radiation in such detail also would help determine the geometrical shape of the universe. The key here is also in the ripples in the radiation. These hot and cold spots of greater and lesser density can only be so large. Each spot must have homogenized into a specific temperature, so all points in the spot must be close enough to exchange heat and radiation. (Think back to the cups of hot chocolate that would be the same temperature only if they were touching.) But radiation (light) can travel only so fast, so there is a maximum size any of these spots could be—about five hundred thousand light-years across. Finding anything in the heavens that we can emphatically state is a specific distance away or a specific distance across is always a big deal. There's not much we can measure directly in the far reaches of space, so knowing that the largest spots are more or less five hundred thousand light-years across is as good as holding up a ruler in the sky.

The trick, then, is to see how long that ruler appears from earth. If it looks like five hundred thousand light-years, then the universe is flat. If it looks larger than that, then the earth is positively curved; smaller, then it's negatively curved. This may not seem inherently obvious at first, but it makes sense when you think about a flat map. As you look at a map of the earth along the highest and lowest latitudes, the image of the land gets severely distorted—the island of Greenland, less than a quarter the size of the United States, looks to be as big as the United States. This is a natural effect of trying to draw something that is positively curved—the spherical earth—onto a flat plane. The same thing happens in space. If we stare at a five-hundred-thousand-light-year ruler way out in the universe, it's going to appear large and distorted if in fact we're living in a spherical universe. It's going to look exactly right only if our universe is flat, if $\Omega = 1$.

At the moment, several less ambitious projects have mapped small regions of the sky with the kinds of detail MAP hopes to achieve. These bits of the background radiation do seem to show an $\Omega = 1$ universe, but MAP will be measuring so much more of the sky that it will, it is hoped, present the definitive description for the shape of the universe.

The incredible details of the cosmic background radiation expected from MAP will offer answers to another question about the shape of space, namely its topology. Topology isn't quite the same thing as whether the universe is flat or curved, rather it has to do with whether the universe is infinite or finite, whether it curves back around onto itself or extends outward forever. Certain geometries of the universe are inextricably linked to certain topologies. A spherical curve will always curve back in on itself into a ball, for example. But flat or saddle-shaped space might curve back in on itself and it might not. (It's hard to imagine that "flat" space could "curve," but it is theoretically possible: imagine a flat sheet of paper on which are drawn several parallel lines, and roll it into a cylinder. Then take the cylinder and curve it into a doughnut. It all seems vaguely curvy to the average eye, but those parallel lines will all still be parallel on this shape, and that's what makes it "flat.")

If you live in a curvy world, then you could look in two directions and theoretically see the same thing from two different sides (much the way that if you could somehow see far enough, while standing on the globe, you could theoretically see the back of your own head). With an incredibly detailed map of the background radiation, we could compare bits in opposite directions to see if they match up. If the patterns overlap perfectly, then we would know that the universe was curving around on itself, that if a spaceship could travel far enough in one direction, it would eventually come back to where it started. This would not automatically tell us whether the universe was flat, but it would tell us whether the universe was finite or infinite, whether it goes on forever or always points you back in the direction you started.

MAP also should give clues as to how much dark matter is thought to be in the universe. While invoking quintessence means that there doesn't have to be quite as much dark matter as previously hoped, almost all aspects of cosmology still require that there be *some.* At a bare minimum, dark matter is still needed to explain why stars and galaxies move so much more quickly than we expect them to—extra gravity produced by some unseen mass is our best guess. Theorists also use dark matter to explain how large bodies such as stars and galaxies could have formed. Atoms and particles just wouldn't have attracted each other enough to form such structures without some extra gravity from dark matter. And lastly, theories about the young universe all insist on some dark matter to make the correct amount of elements we see in the universe today. (Since those first critical formations of helium, hydrogen, and deuterium—known as nucleosynthesis—are one of the main supports for the big bang, getting them right on target is a fairly crucial consideration.)

All of this means that determining how much dark matter there is in the universe, and what that dark matter consists of, still is important to cosmologists. One way to do this is via the cosmic background radiation and what it looked like three hundred thousand years after the big bang. The background radiation began traveling at the moment the universe lost just enough density that light could stream freely instead of getting reabsorbed by some passing particle. Until that point however, all the energy and particles were trapped into one swirling mass of heat. If, in addition to the photons and electrons, that giant ball was made of both baryonic matter (that's all the normal stuff in the universe such as protons and neutrons) and dark matter, then the basic distribution of it all would be different depending on how much of each there was.

The sluggish cold dark matter wouldn't move much. It also wouldn't interact with the photons—after all, if it was affected by light, we'd have been able to *see* it. The baryonic matter, on the other hand, would be flowing around the dark matter like a fluid, exchanging energy with the electrons and the photons. Scientists know enough about the way fluids flow to reasonably predict how these various components would have moved. They even can describe how they should have oscillated: massive waves reverberated through the universe, creating various regions that got denser, or less dense, over time. These kinds of oscillations are nothing more than sound waves, since a sound wave simply changes the density of air or water or matter as it travels through it. These waves have even been referred to as "the music of creation." And, again, sound waves have been studied by scientists for centuries and are fairly well understood.

What does this mean for dark matter? If scientists can get a good look at the patterns of the background radiation, with its patches of dense and less dense areas, and so understand how the acoustic waves were spreading through the early universe, then they can compare those acoustic waves to what they'd expect to see from various amounts of dark matter. This is like hearing the note C played by a flute and played by a tuba—the sound waves that hit your ear are so different that you can tell the instruments apart. In the same way, if you looked at a graph of the sound waves you'd be able to tell the difference. In cosmology, examining the acoustic waves that rumbled through the microwave radiation leads to lovely graphs of frequencies—much like the multiple frequencies that add up to a single booming note on a pipe organ—which can point directly to how much nonbaryonic matter there must have been in that primordial fireball. Cosmologists talk about the "peaks" they expect to see, since the graphs look like S-curves that should show peaks at very specific values for different theories. It is hoped that MAP, with its detailed measurements of the density ripples in the sky, will nail down these peaks once and for all to tell us exactly how much dark matter exists in our universe.

There are some scientists, however, who say that the answer may already be in: smaller experiments may have answered the question and stolen some of MAP's thunder.

BALLOONS AND BOOMERANGS

One of the first experiments to bring back data is known as BOOMERANG (short for Balloon Observations of Millimetric Extragalactic Radiation and Geophysics). The BOOMERANG experiment involved thirty-six scientists from Canada, Italy, the United Kingdom, and the United States. It consisted of a state-of-the-art microwave telescope hanging beneath a balloon that hovered over the Antarctic for some ten days in late 1998. In that short time the telescope could record images from only a tiny portion of space—3 percent of the sky—but it did so at an amazing level of detail, examining fluctuations down to a hundred millionths of a degree.

When the data were analyzed and all the noise filtered out, the BOOMERANG team announced in 2000 that they could clearly see the first peak of the S-curve, the first peak in the predicted acoustic frequencies. This first curve implied that the universe was flat, but didn't help as much with predicting the amount of dark matter in the universe. During the next year, however, they hammered out another two peaks on the curve. The shape of the curve has been largely backed up by several other experiments, including DASI (Degree Angular Scale Interferometer), which also observed the heavens from a perch in Antarctica; MAXIMA (Millimeter Anisotropy Experiment Imaging Array), flown over North America; and CBI (Cosmic Background Imager), flown over Chile.

In the spring of 2001, cosmologists from all four balloon experiments presented findings that seemed to agree with each other. They had hammered down the acoustic frequencies of that early universe. Showing a sound wave that would have produced a tone at the frequency of one cycle every four hundred thousand years—much too low for humans to hear, much too low for even whales to hear—the acoustic peaks implied a flat universe with about 4.5 percent baryonic matter. That leaves the universe with some 30 percent of dark matter; the rest of the universe's budget is used up by dark energy.

The scientists presenting the curves were jubilant, amazed that the data from all four instruments matched up so well, and impressed that they correlated with what theorists had predicted years earlier. Since these acoustic curves match up so nicely with what was predicted for a universe created *à la* big bang and inflation, they lend a great deal of support for current cosmology theories.

OTHER BEGINNINGS

While inflation and its prediction of a perfectly flat universe may have gotten a boost of support from all this heavenly noise, it nevertheless remains true that inflation theory, occurring as it did in the first billionth of a trillionth of a trillionth of a second of the universe's life, is fairly hard to prove conclusively. This means that occasionally new theories about those first moments get proposed.

One such newbie—and perhaps the first truly viable alternative to inflation theory—is called the ekpyrotic universe. ("Ekpyrotic" comes from an ancient Greek term for a universe periodically destroyed and reborn by fire—a possible fate for our own universe in this theory.) The new theory was presented to cosmologists in spring 2001 by Paul Steinhardt, a Princeton physicist previously mentioned in connection with carving out the details of inflation theory, along with his student Justin Khoury and physicists Neil Turok of Cambridge University and Burt Ovrut of the University of Pennsylvania.

The ekpyrotic universe relies on a fairly new branch of physics known as M-theory. M-theory is an extension of comprehensive theories called string theory or supersymmetry theory. No one can quite agree on what the "M" was originally supposed to stand for, but since the theory rests on the idea that particles are made of thin membranes, many people think of it as short for "membrane theory." M-theory exists at the junction between cosmology and particle physics, describing the details of how particles and forces came to exist in the first place, at the beginning of time.

The M-theory world is a fairly complicated place. It's a universe that's made up not of four dimensions—the three dimensions of space and one of time—but eleven. One of these dimensions is time as before, three are the space we experience, one is another spatial dimension we do not experience, and six are rolled into tiny balls, so small that we barely need to concern ourselves with them at all. This is not an intuitive description of our universe, perhaps not one we can ever truly visualize, but when the math for this kind of reality gets worked out, all sorts of fantastic things happen. For one thing, the four forces in our universe—gravity, electromagnetic, strong, and weak—can all be made to fit together into a single theory. Before string theory, gravity was never successfully incorporated into the other three. M-theory also postulates the composition of particles. In early versions of string theory, particles were thought to be strings—tiny little loops vibrating with energy. M-theory added another dimension, claiming that these little loops were actually whole membranes curled into a long tube, which looked like strings when viewed from the end.*

These membranes, known simply as branes, also lie at the heart of the ekpyrotic universe. The pre–big bang universe—or really all of reality beyond our own universe, since in this model such a reality would still exist—was simply cold and vacuous space in which all actions happened slowly and over long periods of time. This is a space with four spatial dimensions and one of time. And in it were floating two large four-dimensional branes, flattened out and parallel to each other, like two sheets hanging on adjacent clotheslines. The entire system likes to keep these branes flat and parallel, and so over a very long time, the branes would align themselves evenly, smoothing themselves out, although some quantum fluctuations on the surface would always remain. One of these flat sheets comprises our visible universe, while the other one is an alternate parallel universe. These two flat sheets would be moving toward each other, gently accelerating until they slammed into each other. The energy of the smash would give the nascent universe all the energy and matter it would have until the end of time; it would be the bang that started our cosmos.

While inflation theory solves the problem of how the universe became flat by theorizing some intensely fast period

---

*Adding this extra dimension to the strings was an important step toward validating string theory. Originally there were several versions of string theory, none of which quite overlapped any of the others. However, once you incorporated an additional dimension, taking the string theories from ten dimensions up to eleven, suddenly all the string theories appeared equivalent, or at least variations on a single theme. It is the eleven-dimensional larger theory, which includes all the previous versions of string theory, that is now known as M-theory.

of growth, the ekpyrotic universe takes the opposite tack. The modern universe is flat because the two branes moved so slowly and had a huge amount of time to even out and flatten. By the time they came into contact, both were already so symmetrical that the universe naturally was, too. Given that amount of time, it's also understandable that the entire brane had reached about the same temperature, that it was symmetrically heated, and thus the horizon problem—the fact that the universe seems to be the same temperature in all directions—is solved as well.

The ekpyrotic universe has answers for several of the other standard problems in the big bang theory.* We see no monopoles in the universe, says the ekpyrotic model, because in this scenario they simply were never produced. By invoking the branes, the universe doesn't ever have to be as hot as it would have been during inflation. Any worrisome bizarre particles that might have been created in that intense heat are instantly eradicated. And last, this universe seeds stars and galaxies and other such large-scale structures simply with the inherent quantum fluctuations that would have been on the primordial branes as they crashed into each other.

Another nice thing about this model is that it's testable—not with currently planned experiments, but nevertheless testable. Inflation theory and the ekpyrotic universe would each have created subtly different versions of the modern universe. For one thing, the gravity waves—waves that rumble through space itself, the way an ocean crest rumbles through water—produced by each would be completely

*These are problems that modern inflation theory solves as well. The point isn't that the ekpyrotic universe surpasses inflation as a theory, only that it seems to be as robust.

different. An inflationary universe would have created gravity waves of all different frequencies, while an ekpyrotic beginning would have created only very specific gravity waves. Two sets of experiments are under way to detect gravity waves, but neither is sensitive enough to detect the different polarizations needed to distinguish between inflationary and ekpyrotic theory.

Another difference between the two is that inflationary theory, for all its successes, doesn't jibe well with superstring physics. As more and more particle physicists embrace M-theory, they may well be drawn over to the ekpyrotic theory for the simple reason that it fits better with the current beliefs about fundamental physics.

At this point it's impossible to predict which theory, if either, will come out on top. The concept of inflation has been around for more than twenty years and is well understood. The ekpyrotic universe, on the other hand, is just getting on its feet. Scientists may soon find some fatal flaw and dismiss it completely, or, as time goes on, more research could show it to fit in perfectly with the rest of cosmology.

While we won't know until such research is done whether the ekpyrotic universe is a viable theory, it's fairly exciting nonetheless. It's the first intriguing attempt to apply M-theory to cosmology theories, and it holds out hope for additional connections between the two. There are other things that M-theory could smooth out within cosmology, if the theories could be perfected. For example, M-theory's extra dimensions could be the answer to our dark matter problem. While we can't sense most things in any parallel universes, M-theory states that gravity could be felt across the dimensions. Perhaps dark matter is merely normal matter in some parallel world.

Multidimensional branes also could reduce the smoothness problem. Our universe might be so homogenous because it's smoothing out in an alternate dimension—regions of the universe are actually close enough in this unseen dimension for them to reach the same temperature, even though they appear too far away in our reality to exchange heat.

Inflation is no doubt still the leading contender for how the universe began. The ekpyrotic model is the new kid on the block, the seventy-five-pound weakling. And, of course, by introducing so many unproved dimensions, it loses the simplicity scientists prefer in their theories. It will have to prove itself in future tests, and other scientists will have to build on it before it will be considered a serious threat. But the model is interesting because it is the first alternative in decades that has captured the mainstream imagination. The ekpyrotic universe stands out because it shows that even though inflation theory has been around and gaining support for twenty-five years, it's not particularly well tested. A new theory like this can still match all the facts as we know them.

7

The Edge of the Unknown

The big bang—glitches and all—offers a satisfactory explanation of how the universe came to be. And yet it leaves so many questions unanswered.

Why is there a universe at all? Is there a creator? How was something formed out of nothing? Why are the laws of physics exactly as they are? Is our universe one among many? How will the universe end? And, of course, what happened before the big bang?

There are a multitude of answers to questions like these, some that even sound plausible. The answers invariably contain a bit of math, a bit of philosophy, and a whole bunch of imaginative license. Basically these questions take us well out of the realm of experimental, provable science. We simply do not have the tools to test any of the theories—and in fact no one has been able to conceive of some future technology that could test most of them. Of course, history has taught that the "inconceivable" often becomes the conceivable faster than anyone could have predicted, so few people would swear that solving such problems is fundamentally out of reach. But for the time being at least, it seems that some questions must be answered by the thoroughly unscientific, time-honored tradition of what makes sense.

Here follow some of the common answers to the questions above, but do not be lured into thinking that they are absolute, agreed-upon, or even reasonably scientific solutions. We have reached the edge of the unknown. It is hard to know how, or even if, it's possible to prove any of them.

WHY ARE THE LAWS OF PHYSICS AS THEY ARE?—PART I

The big bang theory offers an explanation for how the universe started, but it does not inherently satisfy the human longing to truly understand why we exist. We still crave some answer whereby the creation of the universe, the laws of our universe, the birth of mankind, all we see, was somehow inevitable. (Religions seek this as much as science does. A classic Judaeo-Christian theological conundrum is to assess why God, omnipotent and self-sufficient as he is, even bothered to create humans.) Cosmologists seek this sense of inevitability within the bounds of the current mathematical framework.

By postulating gravity, Newton provided this kind of inevitability to describe why the planets traveled around the sun and why they traveled in the perfect ellipses described by Kepler. Theorize a force, understand the way it works and—ta-da!—Newton solved a problem so neatly that everyone agreed the solar system really could work no other way. Cosmologists want that same kind of comfort zone for the big bang, an explanation that neatly, simply, beautifully describes why our universe exists as it currently does, why that first singularity had no choice but to explode into a gigantic expanse of space.

Inflation theory offers a bit of that long-sought sense of inevitability. If inflation theory is correct, it offers a description of the early universe that could have resulted in no other reality than the one we see today. Exponentially expanding space had no choice but to create a smooth, homogenous universe. All well and good, but in many ways inflation theory simply pushes the "whys" back a step, because scientists must next determine why inflation itself was inevitable.

One solution has been postulated by Russian physicist Andrei Linde: chaotic inflation. In this version of events, universes are constantly morphing into existence as part of some larger multiverse. The singularity that started our own universe, then, is a regular occurrence in the multiverse. With one stroke Linde created a model in which the singularity at the beginning of time really isn't that remarkable. Inflation is simply an automatic consequence of a false vacuum. Nothingness inherently leads to somethingness in the theory, the formation of matter becomes a completely expected occurrence, and so we are given a reason why the big bang isn't all that extraordinary.

With so very many of these universes popping into existence, one imagines they must come in all shapes and sizes—or, more accurately, with all different types of physical laws. The force of gravity in an alternate universe might be much stronger than the force of electromagnetism, for example, unlike what exists in our universe. Some of these systems will disappear as quickly as they began—large gravity, of course, would exert a force so strong that a nascent universe would collapse instantly. Another universe might survive but perhaps remain forever a smooth gas, never creating stars and planets; another might be too hot, or too cold, for life. Regardless, with so very many of these universes popping into existence, it isn't all that amazing that at least one was formed as ours was, with just the right conditions for galaxies, for atoms, for intelligent life. With chaotic inflation, Linde created that sense of inevitability we crave.

Except for the obvious question: What started the multiverse? Where did *that* come from? Why does it work the way it does? The multiverse describes a physical reality that has been around forever, and so once again, humans confront whether that idea, the idea of an eternal physical reality with no beginning, is palatable or satisfying on a gut—not scientific—level. (The multiverse as so described is not unlike the steady state theory—an eternally existing universe generating constant creation of new material. And therefore it makes sense that it runs into the same problem: the very human reaction of feeling unsettled when not offered a moment of creation.)

WHY ARE THE LAWS OF PHYSICS AS THEY ARE?—PART II

Another explanation for why the laws of physics turned out exactly as they did comes from M-theory. M-theory is the same theory that underlies the ekpyrotic universe model. It's an example of what's called a Grand Unification Theory (or simply GUT), since one of its main goals is to "unify" the four forces in nature. M-theory holds that all the forces were in fact identical at the very high energies and temperatures of the early universe.

One of the biggest difficulties with M-theory is that the math needed to tackle the physics simply doesn't exist yet. This is a first for science. The usual story goes the other way: As scientists tackle a new field they discover that some esoteric bit of math that no one has much used applies perfectly to the problem at hand. (Einstein, for example, made use of a little-known field of geometry known as Reimann surfaces, named after Bernard Reimann, who developed the math in the nineteenth century.) M-theory has been described as a twenty-first-century science that found its way into the twentieth century—and the math that does apply to it tends to be highly esoteric. Consequently many people discuss strings solely in nice, broad metaphors that are easier to understand. These surely do a disservice to the precision of the theory, but into metaphors we must go.

M-theory attempts to explain how particles and forces were created in the earliest moments. One can liken the process of the four fundamental forces coalescing in the earliest universe to that of ice crystals forming on the surface of a lake. Originally the liquid water atoms are evenly distributed, as likely to face in one direction as in any other. But as the temperature drops in winter, chunks of ice will freeze, forming specific crystal structures, locking water atoms into specific configurations. One can imagine separate chunks forming all over the surface of the lake, each with its water atoms oriented in different positions, and the directions of all the atoms in these various ice floes would have basically been assigned randomly. They would have frozen into that crystalline shape based on random factors—just how cold a wind blew across just which part of the lake at just what time, for example.

M-theory postulates that something similar may have happened to create the forces of nature we see in today's universe. Originally everything was even and the same. The force of electromagnetism equaled gravity equaled the weak force equaled the strong force. As the universe expanded dramatically via inflation, and as temperature and energy levels dropped, the four fundamental forces of nature "froze" into specific strengths. Of course, this version of how the forces came to be still leaves quite a bit to chance, and once again leaves one with as many questions as answers. Since those forces needed to be so very precise for the universe to have lasted as long as it has, one has to wonder how they "randomly" froze into such perfect values. Many scientists hold out hope that as they understand more about branes, they'll discover that the process simply wasn't as random as it currently appears. They hope that the values of the forces are somehow connected, that somehow once one force froze into a specific strength, the others were constrained to specific values—values that would create a universe like the one we live in. In this way the incredible precision of the forces would make sense, and M-theory would have given that missing sense of inevitability to why the laws of physics should be as they are.

Unfortunately, no matter how pretty M-theory is and how well it fits in with cosmology and particle physics, rigorous testing of such Grand Unified Theories is currently well-nigh impossible. The four forces we experience in this universe were connected only at the incredible energies and temperatures right after the big bang. To simulate conditions that extreme would require an accelerator several light-years long.

How Did the Big Bang Create Something out of Nothing?

Some attempts to answer this claim that there really *was* something before the big bang. Linde's chaotic inflation and the ekpyrotic universe both presume some state of reality before our universe was created. But if there truly was only a vacuum before the bang, then inflationary theory has an answer. (With inflationary theory neatly solving so many dilemmas, one begins to understand why cosmologists are reluctant to scrap it, despite its being relatively unprovable.)

First, one needs a reasonable definition of "nothing." The nothingness at the beginning of the universe had zero energy and zero mass, a perfectly acceptable description of a vacuum. Of course, this vacuum was a special quantum vacuum, and quantum mechanics decrees that entities are not "exactly" any number—not even "exactly" zero. So the amount of mass in this scenario actually fluctuates around zero, with bits and particles fleeting into existence for the briefest of moments here and there. Most of these flip instantly back into nothingness.

However, the bit of mass—or field of energy, since mass and energy are equivalent, per Einstein's equation $E = mc^2$—that started our universe came attached to an inflationary field. This field, as you will recall from chapter 4, has an antigravitational force, causing it to expand at unimaginably fast rates. But what's most important for our current discussion is that an antigravitational force has *negative* energy. Negative energy sounds like a neat math trick, simply a number with a minus placed in front of it. But as evidenced by that fantastic repulsive force creating expansion, the concept of negative energy has a very real physical description.

So at the beginning of time there is a huge field of negative energy coupled with the sudden appearance of greater-than-zero mass. But mass and energy are equivalent. Add these two amounts together and you've still got zero total mass and energy. It's possible that if you added up all the energy and all the mass in our entire universe it might still equal zero. So the question "How did something come from nothing?" ceases to be a problem, because there is no "something." To this day we may exist in a zero mass-plus-energy universe, exactly equivalent to the "nothing" that came before the big bang.

WILL THE UNIVERSE END?

In day-to-day life on earth, the past is much easier to discern than the future, but it's not the same with the universe. This book is a testament to the fact that unraveling our past has been fairly complicated. But if we could understand the precise organization of the universe today, we should in principle be able to predict our future fairly accurately.

There are only a few basic possibilities. The universe as a whole could forever expand into an empty wasteland, eventually contract, stay the same forever, or be destroyed in a moment by some mind-bogglingly massive destructive force. None of the scenarios is all that pretty—or one that a human would particularly enjoy experiencing. (And thankfully, with the exception of the last, all of them would be billions of years in the future.)

In the forever-expanding scenario—which would happen if the universe was fundamentally open without enough matter to make it slow down and reverse, as the accelerating-universe theories suggest—all matter would pull farther and farther away over the eons. On shorter scales, though, clumps of matter would pull together. Just as stars have gravitated into galaxies and galaxies have gravitated into clusters over time, so matter would collect into ever denser pockets that would all be ever farther away from each other. These pockets, however, would not be eternal. Stars die, after all, and energy sources run out. As these pockets of light extinguished themselves, blowing up into electrons and protons, the particles would drift even farther apart. And these, too, over unimaginable eons, would decay. Finally, the whole universe would be nothing more than a near vacuum, with an occasional photon or neutrino whizzing by. Still expanding, it would just get more and more dilute . . . forever.

A universe that dies in a fiery crunch, while still disastrous for life, is a bit more exciting. In this scenario, the force of gravity eventually slows the expansion of the universe to a complete halt and then starts pulling it back in, as gravity does to a fly ball, which must inevitably come back down. Initially there would be no obvious sign that the universe had begun to contract, though eventually we would notice that stars seemed to be blueshifted, showing that they're coming toward us, instead of redshifted, like the receding stars we see today. It also has been suggested that somehow our understanding of the direction of time might change as well. At the moment we obviously remember the past and not the future. There seems to be a definite direction to the arrow of time, heading forward. It has been discussed, however, that we may somehow be conditioned to *always* experience time as going away from the initial or final singularity of the universe. So as the universe began to contract, we might start experiencing time backward—remembering the future, while the past remains unknown. This, too, is just the kind of thing that can emerge from math equations but that doesn't have much chance of being tested in the here and now.

Regardless of the time component, things would be fairly different on the return trip. For one thing, while the early universe was naturally very simple and evenly heated, the late universe will have already separated into clusters of galaxies and stars. As these get closer and closer together to one another they will heat up all of space, creating an intense background radiation and making the sky glow red. Hotter and hotter still, space would pass through the colors of the spectrum, getting yellow, than white-hot. Eventually it would be so hot that stars couldn't give off their heat into the atmosphere and would explode from the pressure. In the last few minutes, the heat would keep building up to unimaginable temperatures, stripping all atoms of their particles, leaving nothing but a quark soup. Black holes—the only structures left—would begin to merge, until finally the entire universe as we know it would collapse into one infinitely dense, tiny singularity, ending all space and time.[1]

There is a slight variation on the contracting-universe scenario, in which the universe collapses but bounces back like a yo-yo, recoiling from the final singularity with a new big bang for a new universe. In this model the universe exists forever, expanding and contracting for eternity. It's a fairly comforting vision, if only because it explains how the big bang was created, but from currently understood physics it's fairly hard to accept. As mentioned in chapter 5, there doesn't seem to be any mechanism allowing for a rebound. In addition, if such a bounce had already occurred, we would expect to see heavy particles left over from the universe's previous incarnations, and we don't. Nevertheless, many scientists pursue this model, hoping to find some physical mechanism for this intellectually satisfying solution.

Traditionally, scientists have assumed that expanding forever or collapsing were the two basic options, and if we could just figure out how dense the universe is, we'd be able to know which option would happen. However, the recent discovery that the universe's expansion might be accelerating suggests that we know much less than we think. We've always known that the amount of the universe we can see is less than what actually exists, but an accelerating universe implies it could be thousands and thousands of times bigger than the minute part that's visible.

If that's the case, if we are seeing only a tiny percentage of the whole, then how can we presume to say we know anything? While we might determine the density of our own observable bit, we can't claim to know what is going on throughout the entirety of space—and so could never accurately predict how the universe will end.

That leaves the last two scenarios for the end of the universe: that perhaps it will stay exactly as it is now for all time, or that it might end in some cataclysmic disaster. The former would mean that everything we think we know about the universe, including the fact that it's expanding, is mistaken, and that the laws of physics are not as we understand them. All of that is certainly possible, but not likely. And we have no real way of predicting the latter option either, although in the ekpyrotic universe, in which the universe was created by two membranes colliding, there is certainly a way for such a disaster to occur. All that has to happen is for another floating brane to splat into ours, and all of space would be destroyed. (In fact, in a fairly unsettling version of events it's possible that since the two sheets that formed the universe were accelerating toward each other, the current acceleration of the universe might be a sign that such a great splat of destruction is about to occur again.)

IS THERE A CREATOR?

This question is so squarely outside the realm of science that you'd think no one would even attempt to use science to prove or disprove a creator, but this isn't the case. Many scientists see the very existence of scientific explanations as enough to disprove a deity absolutely. After all, if you can use natural explanations to describe perfectly how the universe began and how it runs, why would you invoke some outside source of power? That would be as ad hoc as supposing that angels are needed to keep the planets spinning around the sun. Gravity replaced angels, and the big bang replaced a creator.

But one can also use science to support a deity. For all the fiddling with inflation theory and string theory in some attempt to explain the inevitableness of modern physics, the fact remains that the universe is finely tuned for human life. The minutest of changes in the size of the weak force would make it impossible for hydrogen—necessary for both the formation of water and the constant fueling of the sun—to form. A change in the strong force would have made it impossible for protons to form at all, or if they did, the change would have forced stars to burn out some quintillion times faster than they currently do. The relative strengths of gravity and electromagnetism had a 1 in 10^{40} chance of being fixed in the exact ratio needed for stable suns to form.[2] These are just a few of the examples of how perfectly the universe seems adjusted for life, and intelligent life, to arise. Many have argued that the coincidences are too great. Such a universe as ours, with stars destined to last as long as needed for a warm planet to form, with water created in abundance, with a gravitational force keeping the universe from collapsing too soon, simply must have been created especially for us.

Neither answer is scientific. Pick the one you like better.

WHAT CAME BEFORE THE BIG BANG?

An elegant answer to this question comes from the Cambridge physicist Roger Penrose, who said that asking this question is like asking what is north of the North Pole. The concept simply doesn't apply. The term "north" no longer has meaning at the Pole, and the term "before" doesn't have meaning when all of space and time was created in the big bang itself. Pre–big bang nothingness is absolute.

Nice rhetoric, but distinctly unsatisfying. Other options as to what came before rely on Linde's multiverse: We're part of some larger conglomeration of universes, continually spawning new growths. Unfortunately, the nature of the big bang implies that we can never see what happened on the other side. This intense singularity destroyed any information we might have reaped from the rest of the multiverse—we can never probe earlier in time.

So that's not such a satisfying an answer either. Neither one really makes you want to lean back and say, "Aha, now I get it. That makes perfect sense." But the solution that makes you feel that way probably won't come from science. Perhaps future generations will come up with an entirely different process for learning such things—something beyond the scientific method itself—but for now any description of a pre–big bang universe brings us right back to storytelling again, no more accurate than ancient religious myths.

The big bang may turn out to be the perfect, hundred-percent-accurate description of how our universe began. Inflation theory, galaxy formation, dark matter, dark energy—these additions may all be hammered out into a perfect, glitch-free picture of cosmology, and yet they will never answer the kinds of questions asked here. Science can explain the "hows," but it will always lead one back to more "whys."

Fundamentally there are only two options to how the universe began: Either it has been around forever, or it began at some point in time. Neither one, philosophically, on a gut level, seems a full answer. And cosmology, as it currently stands, is not equipped to help.

The Future

The history of cosmology is filled with unexamined assumptions that were overturned to gain new perspectives. Some of these assumptions seem odd to us today. Why would anyone have assumed, as Kepler did, that Plato's perfect polygons would govern the way the planets moved? But as we look at modern times, those mistaken assumptions are less obvious. Before Einstein, for example, everyone quite naturally believed that time and space were two separate entities. Einstein saw a connection between them, discarded the old assumption, and bound the two together forever as dependent on each other. Now scientists think of space and time as intricately entwined. The same goes for energy and mass—long perceived to be entirely different, until Einstein said they were equivalent.

As another modern example, take string theory. Everyone assumed we live in a three-dimensional universe. That's a pretty natural assumption, but string theorists now say we may live in an eleven-dimensional universe. Once they discarded their 3-D assumptions, they finally managed to produce theories wherein all four forces were equivalent, something theorists had been trying to do for most of the twentieth century.

The point is that our assumptions are extraordinarily intrinsic. In our lives, space and time don't seem to connect, and our reality exists in only three dimensions, so it's amazing that anyone managed to see beyond these beliefs for the sake of a better theory. It's even more amazing that we can't begin to know what similar assumptions limit us now, what unexamined suppositions about the universe might be roadblocks keeping us from unraveling the next bit of cosmology theory.

We cannot possibly identify all our unquestioned assumptions—this book surely contains many. But cosmology is poised to move quickly over the next few decades, and more of these assumptions will become clear. We will watch as old assumptions are discarded and as new theories take their place.

The big bang theory is robust, well tested, and well accepted, but it is still a theory that is alive, that is changing, that is not quite whole. Parts of it may be overturned completely. With new experiments to be launched, with ever better telescopes, with bigger and bigger accelerators, the years ahead in cosmology are going to be riveting.

A Big Bang Timeline

Following is an approximate timeline of the earliest universe. Though the big bang theory is well accepted by the bulk of physicists as governing the new universe's life from a minute on, various hypotheses have been suggested to tell the story of the first sixty seconds. The numbers and accounts given here should not be considered absolutely accurate.

Planck Time

Time: 10^{-43} second
Universe size: 1/100 cm

At Planck Time, the temperature of the universe is $10^{32\circ}$ Celsius and the density is 10^{90} kg/cm^3 (the density of a rock is roughly 3 kg/cm^3). These numbers are so extreme that they're at the very limits of what we can describe using the theory of relativity. M-theory, the modern, untested theory that represents particles as tiny, vibrating strings, does attempt to describe this era: it was a time when the four fundamental forces were equal (though they would shortly freeze into distinct values). Planck Time is considered the farthest back in history we can probe with current physics. It is, for all intents and purposes, the beginning of our universe.

Inflation

Time: 10^{-36} second to 10^{-33} second
Universe size: Expanding from overwhelmingly small to overwhelmingly large.

If the theory of inflation is correct (see chapter 4), then the entire universe now expands dramatically—to a size perhaps much larger than today's visible universe—in an infinitesimally short time. Inflation produces a dramatically larger universe at earlier times than an "uninflated" big bang.

Alternative: Ekpyrotic Universe

Time: Before the universe's expansion
Universe size: Undetermined

If the ekpyrotic version of the universe is correct (see chapter 6), it would replace the inflationary scenario. Two floating membranes slowly collide, triggering the formation of matter and initiating big bang expansion.

THE FIRST ONE-HUNDREDTH OF A SECOND

Time: 10^{-2} second

Universe size: Variable in different theories. Perhaps about the size of the solar system, perhaps many times larger than the currently observable universe, if certain inflation theories are correct.

During the first one-hundredth of a second, the universe steadily expands but continues to be incredibly dense and incredibly hot. The energies reached could produce the heaviest of particles, not just protons and neutrons, but also uncommon particles such as the family of mesons, rarely observed today except in high-energy accelerators. These continually collide, particles and antiparticles annihilating each other, creating more energy and extra photons—photons instantly reabsorbed by yet more passing particles. Conditions are too extreme for neutrons and protons to ever join up to form atoms.

THE FIRST SECOND

Time: 1 second

The universe has now cooled to below 10 billion °C and the various particles swarming around begin to lose energy. Collisions no longer create enough energy to form new photons; instead, particles and antiparticles simply annihilate each other. Had there been exactly the same amount of matter and antimatter in the universe at this stage, all particles would have been destroyed. Fortunately, there was an excess of particles, and these survived. The ratio of neutrons to protons freezes at about one neutron for every six protons.

Nucleosynthesis

Time: 1 minute

The universe, still expanding, continues to cool until protons and neutrons can stick together via the strong force. They begin to coalesce into the lighter elements such as helium, hydrogen, and deuterium. The first atoms are formed. The bulk of helium and deuterium in the universe today was created during this period of nucleosynthesis. Thus, all theories about the origin of the universe, must correlate to the amount of helium and deuterium we measure today.

Decoupling

Time: 300,000 years

While atoms began to form within a few minutes of the big bang, the universe is nevertheless overwhelmingly dominated by photons (i.e., radiation) for the next several hundred thousand years. This period is consequently known as the radiation era. The photons bounce around, constantly being emitted and reabsorbed by existing particles. As a result, the whole primordial soup has time to reach a fairly even temperature. As the universe continues to expand and cool, the density drops until there is enough room for photons to travel long distances without smashing into other particles. The moment when radiation finally begins to roam free is known as decoupling, and it's this outpouring of perfectly even light that is measured today as the cosmic background radiation. Because the moment of decoupling marks the first time light streamed free, it is also the farthest back in time we can actually "see" with telescopes.

Star and Galaxy Formation

Time: 1 million years

While the universe at the time of the decoupling is incredibly smooth, some slight fluctuations exist. These spots of higher density act as seeds, naturally attracting more material, which coalesces into the cosmic structures we see today. Just how these structures formed is still being debated. More confusingly, the amount of material we see in the universe today does not seem to have enough gravity to create stars and galaxies as we currently observe them. To account for this discrepancy, scientists have postulated that there is nine times more invisible "dark matter" in the universe than matter we can actually see—although the search is still on to find all of it.

Notes

Chapter 1: The First Cosmologies

1. Lucretius, *De Rerum Natura,* trans. R. E. Latham, in *Theories of the Universe: From Babylonian Myth to Modern Science,* ed. Milton K. Munitz (New York: The Free Press, 1957), p. 56. Originally published by Penguin Books (Baltimore, 1951).

Chapter 2: The Birth of the Big Bang

1. George Gale and John R. Urani, "Philosophical Aspects of Cosmology," in *Cosmology: Historical, Literary, Philosophical, Religious, and Scientific Perspectives,* ed. Noriss S. Hetherington (New York: Garland, 1993), p. 561.

Chapter 3: The Search for Proof

1. Based on a discussion with Helge Kragh, 2000.
2. Helge Kragh, *Cosmology and Controversy: The Historical Development of Two Theories of the Universe* (Princeton, N.J.: Princeton University Press, 1996), p. 180.

Interlude: Popular Reactions

1. Helge Kragh, *Cosmology and Controversy: The Historical Development of Two Theories of the Universe* (Princeton, N.J.: Princeton University Press, 1996), p. 244.
2. Patrick A. Wilson, "The Anthropic Principle," in *Cosmology: Historical, Literary, Philosophical, Religious, and Scientific Perspectives,* ed. Noriss S. Hetherington. (New York: Garland, 1993), p. 507.
3. John Leslie, *Universes* (London: Routledge, 1996), p. 4.
4. Ibid., pp. 13–14.

Chapter 4: Scientific Reactions

1. Alan Guth, *The Inflationary Universe: The Quest for a New Theory of Cosmic Origins* (Reading, Mass.: Helix Books, 1997), pp. 22–24.
2. Ibid., pp. 173–175.
3. Paul Davies, *The Last Three Minutes: Conjectures about the Ultimate Fate of the Universe* (New York: Basic Books, 1994), p. 71.
4. Interview with Christopher Stubbs, 1993.
5. Marcus Chown, *Afterglow of Creation: From the Fireball to the Discovery of Cosmic Ripples* (London: Arrow Books, 1993), p. 113.
6. Ibid., p. 137.
7. Joseph Silk, *The Big Bang,* 3rd ed. (New York: Freeman, 2001), pp. 172, 174.

Chapter 5: Glitches

1. Joseph Silk, *The Big Bang,* 3rd ed. (New York: Freeman, 2001), p. 401.

2. John Maddox, *What Remains to Be Discovered: Mapping the Secrets of the Universe, the Origins of Life, and the Future of the Human Race.* (New York: Touchstone, 1999), p. 46.

3. Geoffrey Burbridge, Fred Hoyle, and Jayant V. Narlikar, "A Different Approach to Cosmology," *Physics Today* (April 1999), pp. 38–44.

4. J. Silk, *The Big Bang,* p. 396.

Chapter 7: The Edge of the Unknown

1. Paul Davies, *The Last Three Minutes: Conjectures about the Ultimate Fate of the Universe.* (New York: Basic Books, 1994), pp. 119–123.

2. John Leslie, "The Anthropic Principle Today," in *Modern Cosmology and Philosophy,* ed. John Leslie. (Amherst, N.Y.: Prometheus Books, 1998), p. 292.

Glossary

anthropic principle. The notion that since humans exist in the universe, the universe must be a place conducive to human life. A stronger version of the principle maintains that *any* universe would necessitate the existence of humans.

apparent brightness. A measure of how bright a star appears to be when viewed from Earth.

atom. The smallest part of a substance that can exist and still be that substance. An atom is made of a nucleus surrounded by electrons. Atoms make up elements; the *kind* of element is dependant on the number of electrons, protons, and neutrons present in the atom. When you break down atoms further, the parts cease to be elements and are considered subatomic particles.

big bang theory. The widely held belief that the universe began approximately fifteen billion years ago, when a fireball expanded into the universe we see today. The three main pieces of evidence for the theory are the cosmic background radiation, nucleosynthesis, and the expansion of the universe.

big crunch. A friendly name for the decidedly unfriendly scenario whereby the universe ceases to expand, reverses direction, and eventually collapses in on itself.

branes. The most basic form of matter in M-theory, an unproven theory describing the fundamental forces and particles in the universe. A brane with no dimensions is a point-like particle, a brane with one dimension is a string, and a brane with two dimensions is a sheet.

cepheid variables. A class of stars whose brightness varies in predictable patterns over time. Scientists can use the rate of these changes to determine the absolute brightness of the star. Then, by comparing it to how bright the star appears on Earth—the apparent brightness—they can figure out how far away the Cepheid is. Cepheids, therefore, are used to accurately measure distances around the universe.

cosmic background radiation (or cosmic microwave background). Radiation left over from the big bang—specifically about 300,000 years after the big bang, when the baby universe was finally cool enough for radiation to stream freely. That this radiation is still detectable today is considered one of the proofs for the big bang theory.

cosmological constant. A number arbitrarily included by Einstein in the first relativity equations. The constant kept the universe static, a counterbalance to the force of gravity. While Einstein later believed the cosmological constant to have been a mistake, modern cosmologists have begun once again to include it, or something similar, based on recent evidence that the universe may in fact have some kind of antigravitational force causing it to expand faster than expected.

cosmology. The study of the universe. Cosmologists seek to learn how the universe formed, what it looks like today, and what its ultimate fate may be.

curvature of space. Refers to the concept that all of space is curved due to the matter it contains. One can envision this by thinking of our globe: the earth's gravity keeps one walking along a curved line, despite the fact that the path may appear flat to the walker. Space may be similarly curved, though it appears flat.

dark matter. Any matter that we cannot see, either via visible light or other electromagnetic radiation. Dark matter is detectable solely because we observe its gravity. The amount of dark matter in the universe will affect whether the universe itself will ever stop expanding and reverse into a big crunch.

Doppler effect. The change in frequency in a wave that an observer would note when the object emitting the wave is moving. One can hear the effect in the changed pitch of an ambulance siren as it passes. The Doppler effect also changes the frequency of light coming from a moving star (and per the expansion of the big bang theory, all stars are moving), changing the color of the light.

ekpyrotic universe theory. A new and untested theory based on M-theory that suggests the universe began when two gigantic branes collided. This theory does not contradict the big bang theory, but, if true, would likely replace the inflation theory.

electron. A subatomic particle found in an atom, orbiting the nucleus. In the early years of the universe, however, electrons roamed freely, apart from any atoms. Electrons are much smaller than protons and neutrons, and are negatively charged.

elementary particles. Particles that cannot be broken into smaller pieces. These include, among others, quarks and electrons. An atom is not an elementary particle since it can be broken down even further. The big bang theory postulates that all elementary particles in the universe today were made in the first moments of the universe's existence.

expansion. Discovered by Edwin Hubble in 1929, the observation that the bulk of matter in the universe seems to be moving away from us, implying that the universe is growing. Expansion is considered one of the proofs for the big bang theory.

galaxy. A collection of stars. Most galaxies are so far away that while a single galaxy may contain ten billion stars, it appears as a single star to the naked eye.

gravity. The mutual attraction all matter has for all other matter. Sir Isaac Newton perceived this as a force pulling two objects together. With relativity, Albert Einstein suggested gravity was caused by matter curving the space around it, causing objects to fall inward, much like a weight sitting in the middle of a trampoline that causes everything else to slip into the center.

inflation theory. A theory proposed by Alan Guth in 1980 stating that the earliest moments of the universe were characterized by dramatic, exponential rates of expansion. While not proven, it fits in with the big bang theory and is accepted by much of the cosmological community.

light year. The distance light travels in a year, equal to 10,000,000,000,000 km. A light year is not a measure of time.

MACHO. Massive Compact Halo Object, which is a dense celestial object the size of a large planet, such as Jupiter. MACHOs are the only form of dark matter that has been detected.

M-theory. An untested theory that seeks to explain the basic structure of matter and how the four forces of nature—gravity, electromagnetism, the weak electric force, and the strong electric force—are linked together. M-theory is the newest name for what was once called superstring theory.

multiverse. A collection of numerous universes in which each universe is constantly spawning new ones in bursts of inflation much like the burst that is thought to have created our own. The multiverse scenario solves the problem of why our universe is so finely tuned for life: in a multiverse, ours is one of many, and it is not so odd that one just happened to have the correct conditions for humans to exist.

neutrino. An extremely light or weightless particle. If neutrinos are discovered to have mass, then they may be a candidate for dark matter.

neutron. A subatomic particle found in the nucleus of an atom. The big bang theory states that before the universe cooled enough to allow atoms to form, neutrons traveled freely. A neutron is about the size of a proton but has no electric charge.

nucleosynthesis. According to the big bang theory, the period of time during which all the light elements that exist today—hydrogen, helium, deuterium—were created. Since the theory predicts amounts of these elements that coincide with what scientists observe, nucleosynthesis is one of the proofs for the big bang theory.

nucleus. Lying at the center of an atom, a nucleus is made up of protons and neutrons. Different amounts of protons and neutrons correspond to different elements.

omega (Ω). The Greek letter is used to represent a number that corresponds to the shape of the universe. If omega equals one, as many theorists believe, the universe is flat (and, in the most basic scenario, will expand until it comes to a slow halt). If omega is less than one, the universe is shaped like a saddle (and may expand forever). If omega is greater than one, the universe curves in like a sphere (and may eventually stop expanding to collapse into a big crunch). The shape of the universe depends on several factors including how much matter is in the universe (so whether or not there is dark matter would have an effect) and whether there is a cosmological constant counteracting the force of gravity.

photon. An elementary particle that is the fundamental particle of light.

planets. Large bodies—like Earth, Mars, Venus, and so forth—that orbit stars. Planets look like stars to the naked eye; however, their position moves from night to night, which stars never do.

proton. A subatomic particle found in the nucleus of an atom. The big bang theory states that before the universe cooled enough to allow atoms to form, protons traveled freely. The proton is the same size as a neutron and has a positive charge.

quasar. Short for quasi-stellar object, huge bodies that emit tremendous amounts of energy. Astronomers do not yet have a full understanding of them, though they are believed to be extremely distant and representative of the universe at a young age.

redshift. The Doppler effect that takes place when galaxies and stars move away from us, making their light appear redder than it actually is. A star's redshift is thought to correspond to its speed, distance, and age.

steady state hypothesis. A hypothesis that was popular in the 1950s that suggested the universe has always appeared throughout time just as it does now, and that there was no initial moment of creation as suggested by the big bang theory.

supernova. Occurring during the final days of a star's life, a violent explosion so bright that it can sometimes be seen from Earth during the day. Scientists know how to measure the distance to supernovae, making them an effective (if rare) tool in determining the size and shape of the universe.

WIMP. Short for Weakly Interacting Massive Particle, this exotic class of particle is a theoretical candidate for dark matter. No proof of the existence of WIMPs has been found, however.

Bibliography

Barrow, John D. *Impossibility: The Limits of Science and the Science of Limits.* Oxford: Oxford University Press, 1998.

Bartusiak, Marcia. *Through a Universe Darkly: A Cosmic Tale of Ancient Ethers, Dark Matter, and the Fate of the Universe.* New York: HarperCollins, 1993.

Chown, Marcus. *Afterglow of Creation: From the Fireball to the Discovery of Cosmic Ripples.* London: Arrow Books, 1993.

Craig, William Lane, and Quentin Smith. *Theism, Atheism, and Big Bang Cosmology.* Oxford: Clarendon Press, 1995.

Davies, Paul. *The Last Three Minutes: Conjectures about the Ultimate Fate of the Universe.* New York: Basic Books, 1994.

Ellis, George. *Before the Beginning: Cosmology Explained.* London: Boyars/Bowerdean, 1994.

Ferris, Timothy. *Coming of Age in the Milky Way.* New York: Morrow, 1988.

———. *The Whole Shebang: A State-of-the-Universe(s) Report.* New York: Simon & Schuster, 1997.

Gleiser, Marcelo. *The Dancing Universe: From Creation Myths to the Big Bang.* New York: Dutton, 1997.

Goldsmith, Donald. *The Runaway Universe: The Race to Find the Future of the Cosmos.* Reading, Mass.: Helix Books, 1999.

Greenstein, George. *Portraits of Discovery: Profiles in Scientific Genius.* New York: Wiley, 1997.

Guth, Alan. *The Inflationary Universe: The Quest for a New Theory of Cosmic Origins.* Reading, Mass.: Helix Books, 1997.

Harrison, Edward. *Darkness at Night: A Riddle of the Universe.* Cambridge, Mass.: Harvard University Press, 1987.

Hellman, Hal. *Great Feuds in Science: Ten of the Liveliest Disputes Ever.* New York: Wiley, 1998.

Hetherington, Noriss S., ed. *Cosmology: Historical, Literary, Philosophical, Religious, and Scientific Perspectives.* Garland Reference Library of the Humanities, vol. 1634. New York: Garland, 1993.

Hoyle, F., G. Burbidge, and J. V. Narlikar. *A Different Approach to Cosmology: From a Static Universe through the Big Bang Towards Reality.* Cambridge, Eng.: Cambridge University Press, 2000.

Kolb, Rocky. *Blind Watchers of the Sky: The People and Ideas That Shaped Our View of the Universe.* Reading, Mass.: Helix Books, 1996.

Kragh, Helge. *Cosmology and Controversy: The Historical Development of Two Theories of the Universe.* Princeton, N.J.: Princeton University Press, 1996.

Krauss, Lawrence. *Quintessence: The Mystery of Missing Mass in the Universe,* second ed. New York: Basic Books, 2000.

Leslie, John. *Universes.* London: Routledge, 1996.

———, ed. *Modern Cosmology and Philosophy.* Amherst, N.Y.: Prometheus Books, 1998.

Lindberg, David C., and Ronald L. Numbers, eds. *God and Nature: Historical Essays on the Encounter between Christianity and Science.* Berkeley: University of California Press, 1986.

Livio, Mario. *The Accelerating Universe: Infinite Expansion, the Cosmological Constant, and the Beauty of the Cosmos.* New York: Wiley, 2000.

Maddox, John. *What Remains to Be Discovered: Mapping the Secrets of the Universe, the Origins of Life, and the Future of the Human Race.* New York: Touchstone, 1999.

Mitchell, William C. *The Cult of the Big Bang: Was There a Bang?* Carson City, Nevada: Cosmic Sense Books, 1995.

Munitz, Milton K., ed. *Theories of the Universe: From Babylonian Myth to Modern Science.* New York: The Free Press, 1957.

North, John. *The Norton History of Astronomy and Cosmology.* New York: Norton, 1994.

Rees, Martin. *Before the Beginning: Our Universe and Others.* Reading, Mass.: Helix Books, 1997.

Rothman, Tony, and George Sudarshan. *Doubt and Certainty.* Reading, Mass.: Helix Books, 1998.

Shlain, Leonard. *Art and Physics: Parallel Visions in Space, Time, and Light.* New York: Morrow, 1991.

Silk, Joseph. *The Big Bang,* third ed. New York: Freeman, 2001.

Sproul, Barbara. *Primal Myths: Creation Myths Around the World.* New York: HarperCollins, 1991.

Weinberg, Steven. *The First Three Minutes: A Modern View of the Origin of the Universe,* second ed. New York: Basic Books, 1993.

Index

acceleration. *See also* expanding universe
  expanding universe, 129–132, 140
  quintessence, 144
  supernovas, 141–142
air
  Aristotle, 19
  Plato, 15
Alcock, Charles, 108
Alfonso X (king of Spain), 4
Alfven, Hannes, 133
Alpha Centauri, 49
Alpher, Ralph, 60–61, 79
Amyntas II (king of Macedon), 18
Anaximander, 14
Andromeda, 49, 50
anthropic principle, 88–90
antimatter, 134
antiquity, philosophy, 3, 4 *See also* Greece (ancient)
argument from fine tuning concept, 85–86
Aristotle, 4, 5, 14, 15–20, 22, 24, 25, 28, 30, 33, 63, 116
Arp, Halton, 127, 128
astronomy
  background radiation studies, 111–115
  Brahe, 23–26
  cosmology and, 37
  dark matter, 106–110
  distance determination, 127–128
  history of, 37–38
  Hubble, 48–52
  Kepler, 26–28
  radio astronomy, 72–73
  steady state theory, 70
  telescope, 35
atomic nucleus, 57, 59
atomic structure, 58–59

Babylonian myths, 3, 9–10
background radiation
  balloon antennae studies, 111, 125
  big bang theory support, 120, 121, 134–135
  detection of, 62–63, 65, 76–80, 110
  Microwave Anisotropy Probe (MAP), 105n, 125–126, 145–150
  plasma model, 133–134
  quasi-steady state theory, 132–133
  steady state theory, 120
  temperature of, 62, 77, 97, 113
balloon antennae studies, background radiation, 111, 125, 150–152
Balloon Observations of Millimetric Extragalactic Radiation and Geophysics (BOOMERANG), 150–152
baryonic matter, 149, 152
beauty must be divine concept, 84
Betelgeuse, 49
Bethe, Hans, 59, 60, 61, 68

big bang theory, 34. *See also* cosmology; universe
- acceptance of, 93, 119
- alternate interpretations of, 126–129
- alternatives to, 132–135
- background radiation, 76–80, 112–113
- flatness problem, 104–110
- future research in, 137–157
- galaxy formation problem, 110–115
- history of, 6, 37, 45
- horizon problem, 97–104
- particle physics, 58–62, 74–75
- popular reactions to, 81–90
- reinterpretation of, 1–2
- religion and, 63
- science and, 10–11, 80
- statement of, 9
- steady state theory contrasted, 63–70
- supporting facts, 120–122
- term of, 65–66
- timeline, 175–179
- uncertainties in, 116–117, 122–126, 159
- universe shape, 72

black holes, 108, 115, 121, 132
Bohr, Niels, 57
Bondi, Hermann, 64, 66, 70
BOOMERANG (Balloon Observations of Millimetric Extragalactic Radiation and Geophysics), 150–152
Brahe, Tycho, 21–22, 23–26, 28, 31, 140
branes, M-theory, 164
Buddhism, big bang theory, 86–87
Burbridge, Geoffrey, 74, 132, 134–135
Burbridge, Margaret, 74
Burke, Bernie, 78

calculus, 32
Carter, Brandon, 88
Cassiopeia, 24
Catholic Church. *See also* religion
- big bang theory, 83
- Galileo, 30

CBI (Cosmic Background Imager), 151
Cepheid stars, 49–50, 67, 127, 141
chaotic inflation theory, 104, 161–162, 165
COBE (Cosmic Background Explorer), 112–115, 116, 117, 124–125, 145
cold dark matter, 114–115, 149
comets, 24, 25
complexity problem, future research, 138–142
contracting/expanding universe, 134, 169
contracting universe, 168–169
Copernicus, 5, 22–23, 26, 30, 31, 35, 66, 81, 105
Cosmic Background Explorer (COBE), 112–115, 116, 117, 124–125, 145
Cosmic Background Imager (CBI), 151
cosmological constant, 44, 46, 85, 130, 131, 139, 140, 143
cosmology. *See also* big bang theory; universe
- acceptance of, 93, 100
- future of, 173–174

gravity, 34
Greece (ancient), 3, 12–21
modern, 37, 58
philosophy, 3, 4
religion, 3, 55, 63, 83–88
research methods, 137
science, 3–4, 52–53, 79–80
subject of, 2
creation
big bang theory, 59–62, 65
myths of, 3, 9–10
primordial state, 69
religion, 83–84, 170–171

dark energy, future research, 142–145
dark matter
big bang theory uncertainty, 122–123, 124
evidence of, 106–110
gravity, 114–115, 116
Microwave Anisotropy Probe (MAP), 148–150
M-theory, 156–157
DASI (Degree Angular Scale Interferometer), 151
decoupling, 178–179
Degree Angular Scale Interferometer (DASI), 151
deity. *See* religion
Demiurge, 15
de Sitter, Willem, 45, 56
deuterium, 60
big bang theory support, 120–121
creation of, 74–75
Dicke, Bob, 78–79
Digges, Thomas, 35
Dingle, Herbert, 53
distance, velocity and, 51
Doppler, Christian Johann, 38
Doppler effect, expanding universe, 126

earth
age of, 65
Aristotle, 19
Brahe, 25
Copernicus, 66
Galileo, 30
Kepler, 27
Plato, 15
shape of, 71
Eastern philosophy, big bang theory, 86–87
Einstein, Albert, 1, 34, 38–47, 64, 166, 173
big bang theory and, 65
General Theory of Relativity, 41–43
gravity, 55, 106
Reimann surfaces, 163
Special Theory of Relativity, 39–40
universe, 43–47, 51, 72, 105, 130, 131, 139
Einstein, Lieserl, 39
Einstein, Mileva Maric, 38–39
ekpyrotic universe, 152–157, 162–165, 170, 176
electromagnetism, Grand Unified Theories, 100
electrons, atomic structure, 58–59
elements
Aristotle, 19
background radiation, 78–79
big bang theory, 58, 68
Plato, 15
quintessence, 143–145
stars, 74

ellipses
  Galileo, 29
  Kepler, 27
energy
  flux density, 72
  Higgs Field, 101
  negative energy, 166
  ylem, 61–62
Enuma Elish (creation myth), 9–10
epicycle, 20
ether, Aristotle, 19
Eudoxus, 16–17
European Space Agency, 126
expanding/contracting universe, 134, 169
expanding universe
  acceleration, 129–132
  acceptance of, 63, 66, 94
  big bang theory, 120, 121, 126–129
  calculations, 95–96
  evidence for, 51–52
  speed of, 123–124
  theory of relativity, 44–47

fine tuning concept, argument from, 85–86
fire
  Aristotle, 19
  Plato, 15
flatness problem, big bang theory, 104–110
flux density of energy, measurement of, 72–73
Fowler, Willy, 74
Fraunhofer, Joseph, 38
frequency, 38
Friedmann, Aleksander, 45–46, 56, 57, 139
fundamentalism, religious reactions, 82
future research, 137–157
  BOOMERANG, 150–152
  complexity problem, 138–142
  dark energy, 142–145
  ekpyrotic universe, 152–157
  Microwave Anisotropy Probe (MAP), 145–150

galaxies
  age of, 68
  dark matter, 106–110
  distance of, 67, 68
  formation of, 110–115, 124–125, 179
  quasars, 127–128
  speed of, 50–51, 123
  spiral nebulae, 50
  steady state theory, 66
Galileo Galilei, 5, 22, 28–30, 31, 32, 37, 41, 81, 83
Gamow, George, 56–63, 65, 66, 68, 76, 78, 79, 96
gas clouds, 115
General Theory of Relativity, statement of, 41–43. *See also* relativity theory
geometry
  astronomy, 49
  Kepler, 27
  Reimann surfaces, 163
God. *See* religion
Gold, Thomas, 64, 66, 70
gold atom, 59
Grand Unified Theories (GUT), 100
  dark matter, 108
  M-theory, 162–165

gravity, 85. *See also* cosmological constant
  black holes, 132
  chaotic inflation theory, 161–162
  dark matter, 106, 114, 116
  Galileo, 29
  Grand Unified Theories, 100
  Newton, 32–34, 136, 160
  relativity theory, 41–43, 72
  resistance to, 1
Greece (ancient), 3, 12–21, 31, 143
GUT (Grand Unified Theories), 100
  dark matter, 108
  M-theory, 162–165
Guth, Alan, 99–103, 105

Hawkings, Stephen, 113
heaven, Aristotle, 19–20
Heisenberg, Werner Karl, 57
helium, 60
  big bang theory support, 120
  creation of, 74–75
Herman, Robert, 60, 61, 79
Herschel, John, 37–38
Herschel, William, 37–38
Higgs Field, 100–101
horizon problem, big bang theory, 97–104, 124
hot dark matter, 114–115
Hoyle, Fred, 64, 65, 66, 70, 74, 75, 84, 120, 121, 132, 133, 135
Hubble, Edwin, 47, 48–52, 64, 67, 127, 128, 129, 141
Hubble's Law, 51, 64, 141
Hubble Space Telescope, 6
hydrogen atom, 59

inertia, 40
inflation theory, 100–104, 105, 110, 116, 124, 131, 143, 152, 154, 156, 157, 161, 165–166, 176

Jeans, James, 52–53
Judaism, big bang theory, 86–87
Jupiter, 27

Kepler, Johannes, 5, 21–22, 23, 26–28, 29, 31, 32, 33, 83, 140, 160, 173
Khoury, Justin, 153
Kuhn, Thomas, 11

Leavitt, Henrietta, 49
Lemaître, Georges, 46–47, 56, 83
Leslie, John, 89
light
  Newton, 32
  radio astronomy, 72–73
Linde, Andrei, 103, 104, 161, 162, 165, 172
literature, big bang theory, 86–88
lithium, creation of, 74
Lovell, Bernard, 84
Lucretius, 14
Luther, Martin, 30
Lutheranism, 26, 83

MACHOs (Massive Compact Halo Objects), 107, 108–109
Mars
  Copernicus, 23
  Kepler, 21–22, 27, 28
  Ptolemy, 21
mass, theory of relativity, 43–44
Massive Compact Halo Objects (MACHOs), 107, 108–109

Mastlin, Michael, 26
mathematics
  cosmological constant, 143
  cosmology and, 52–53
  Einstein, 44
  Grand Unified Theory, 100–103
  Greek philosophy, 17, 20
  M-theory, 163
  Newton, 32, 33
  quantum mechanics, 57
  reality and, 47
  science, 87
  theory and, 64
Mather, John, 112
matter
  antimatter, 134
  background radiation, 78–79
  steady state theory, 66
  theory of relativity, 72
MAXIMA (Millimeter Anisotropy Experiment Imaging Array), 151
Mercury, 27, 41
microlensing, starlight, 108–109
Microwave Anisotropy Probe (MAP), 105n, 125–126, 145–150
microwave background radiation. *See* background radiation
Middle Ages, religion, 19, 20
Millimeter Anisotropy Experiment Imaging Array (MAXIMA), 151
Milne, E. A., 52–53
momentum, Galileo, 29
moon, 38
  Aristotle, 19
  Galileo, 29
  gravity, 33
  observation, 12
morals, Socrates, 15
M-theory
  ekpyrotic universe, 153–157
  laws of physics, 162–165
  Planck Time, 175
multiverse, chaotic inflation theory, 161–162, 172

Narlikar, Jayant, 132, 133, 135
National Aeronautics and Space Administration (NASA), 4, 24, 105n, 112, 125
Native American myths, 3
nature, religion and, 14
nebulae, 48–49, 50
negative energy, 166
Neptune, 106
neutrinos, 4, 107–108, 114
neutrons
  atomic structure, 58–59
  big bang theory, 60–61
Newton, Isaac, 31–35, 40, 41, 42, 43, 116, 136, 160
nothing, 165–166. *See also* vacuum
nucleosynthesis, 178

observation
  Aristotle, 18
  Brahe, 24–25
  Galileo, 29
  heavens, 12, 13
  Plato, 15
  Ptolemy, 20–21
  science and, 53, 55
Occam's Razor, Copernicus, 22–23
open-mindedness, theory and, 2, 5, 11
optics, 38
orbits

gravity, 33
Ptolemy, 20–21
Ovrut, Burt, 153

paradigm shifts, theory, 11
particle physics
big bang theory, 58–62, 74–75
dark matter, 108
Peebles, Jim, 78
Peloponnesian War, 15
pendulum, Galileo, 28–29
Penrose, Roger, 172
Penzias, Arno, 76–78, 79, 80
perfect cosmological principle, 66
perfections, 15
philosophy
big bang theory, 88–90
cosmology, 3, 4, 5
Greece (ancient), 3, 12–21, 31
religion, 84–85, 90
science and, 90
theory and, 11–12, 52–53
photons, decoupling, 178–179
Pius XII (pope of Rome), 63
Planck, Max, 11
PLANCK instrument, 126
Planck Time, 175–176
planets, 38
Aristotle, 18, 20
Galileo, 29–30
gravity, 33
Kepler, 21–22, 27–28
observation, 13
Ptolemy, 20–21
plasma theory, 133–134
Plato, 3, 14, 15–17, 27, 31, 63, 173
politics, Soviet Union, 57
popular reactions
philosophy, 88–90
religion, 82–88
to theory, 81–82
primeval atom concept, 46–47
primordial state, creation, 69
prism, 38
protons, atomic structure, 58–59
Ptolemy, 20–21, 22, 23
Pythagoreans, 22

quantum mechanics, 57
quasars
described, 73–74
galaxies, 127–128
quintessence, 143–145

radar research, 64
radiation, ylem, 62
radioactivity, 81
radio astronomy, 72–73, 76–77
reality, mathematics, 17, 47
rebounding universe, 134, 169
redshifts
expanding universe, 127–129
quasars, 73–74
starlight, 50, 51
supernovas, 141–142
Rees, Martin, 85, 86
Reimann, Bernard, 163
Reimann surfaces, 163
relativity theory, 64, 173
big bang theory and, 65
General Theory of Relativity, 41–43
reinterpretation of, 1
Special Theory of Relativity, 39–40
speed of light, 127

relativity theory, (continued)
  universe, 43–47, 72, 105–106, 130, 131, 139
religion
  background radiation, 113
  big bang theory and, 63, 82–88
  cosmology, 3, 5, 14, 55
  existence and, 160
  expanding universe, 46
  Galileo, 30
  Lutheranism, 26
  Middle Ages, 19, 20
  Newton, 34–35
  relativity theory, 42
  science, 6, 34–35, 82–88, 90, 170–171
  Soviet Union, 57
retrograde, 20
Rheticus, 23
Roll, Peter, 78, 79, 80
Rubin, Vera, 106
Rudolph II (Holy Roman Emperor), 26
Rutherford, Ernest, 57
Ryle, Martin, 73

Sandage, Allan, 68, 70
satellites, background radiation studies, 111–115
Saturn, 35
  Galileo, 29
  Kepler, 27
Schrödinger, Erwin, 57
science
  big bang theory, 10–11
  cosmology, 3–4, 52–53, 55, 79–80
  religion, 6, 34–35, 82–88, 90, 170–171
  Soviet Union, 57
  theory, 2, 11–12
shape-of-space concept, 95
simplicity concept
  big bang theory, complexity problem, 138–142
  Copernicus, 22–23
  inflation theory, 116
Simplicius, 16n
Smoot, George, 113
Socrates, 15
solar eclipse, 42
solar system. *See also* sun
  Brahe, 25
  composition of, 3–5
  Kepler, 27–28
Soviet Union, 57
Special Theory of Relativity, statement of, 39–40. *See also* relativity theory
spheres, Aristotle, 18, 19–20, 25
spiral nebulae, 48–49, 50
standard candle, supernovas, 141
starlight
  background radiation, 135
  deflection of, 42
  measurement of, 49–50
  microlensing, 108–109
  redshifts, 50
  speed of, 45
stars
  age of, 67, 122
  Brahe, 23–26
  distance of, 127–128
  elements, 59–60, 74
  flux density of energy, measurement of, 72–73
  formation of, 114–115, 179
  Galileo, 29–30
  observation, 12–13
  speed of, 38, 123

steady state theory
  challenges to, 67–70, 73–75, 120
  development of, 63–66
  quasi-steady state, 132–133
  religion, 84
  universe shape, 72
Stebbins, Joel, 68
Steinhardt, Paul, 102, 103, 144, 145, 152
string theory, 173–174
strong force, atomic structure, 59, 85
sun. *See also* solar system
  Brahe, 25
  Copernicus, 22
  Galileo, 29, 30
  Kepler, 26–27
  observation, 12, 13
  Ptolemy, 20–21
Super-Nova/Acceleration Probe (SNAP), 145
supernovas
  observation of, 24–25, 140
  speed of, 1
  studies of, 141–142

telescope, 5, 25, 41
  astronomy, 35
  Galileo, 29, 37
  Hubble Space Telescope, 6
temperature
  background radiation, 62, 77, 97, 113–114
  contracting universe, 168–169
  dark matter, 114
  horizon problem, 97–104
  Planck Time, 175
  thermal equilibrium, ylem, 61–62
theory
  mathematics and, 64
  open-mindedness and, 2, 5
  paradigm shifts, 11
  public consciousness of, 81–82. *See also* popular reactions
  reality, 17
  science and, 11–12, 52–53
thermal equilibrium, ylem, 61–62
time, contracting universe, 168
timeline, 175–179
  decoupling, 178–179
  ekpyrotic universe, 176
  first second, 177–178
  inflation, 176
  nucleosynthesis, 178
  Planck Time, 175–176
  star and galaxy formation, 179
Tolman, R. C., 53
Turner, Ken, 78
Turok, Neil, 153
Tye, Henry, 100

universe. *See also* cosmology; expanding universe
  age of, 64–65, 67, 122
  composition of, 3–5
  contraction of, 168–169
  density of, 95, 122–124
  ekpyrotic universe, 152–157, 162–165, 170, 176
  end of, 166–170
  multiverse, 161–162, 172
  rebounding universe, 134, 169
  shape of, 70–73, 85, 95, 102–103, 104–110, 123–124, 138–140, 147–148
  size of, 67
  speed of, 1
  theory of relativity, 43–47

Uraniborg, 24, 25
Uranus, 106

vacuum, 165
  acceleration of expanding universe, 130
  Higgs Field, 101, 103
velocity, distance and, 51
Venus, 4
  Galileo, 29–30
  Kepler, 27

water
  Aristotle, 19
  Plato, 15
wavelengths, 38
Weakly Interacting Massive Particles (WIMPs), 107–108, 109, 114
white dwarfs, 108, 109
Whitford, Albert, 68
Wilkinson, Dave, 78, 79, 80
Wilson, Robert, 76–78, 79, 80
WIMPs (Weakly Interacting Massive Particles), 107–108, 109, 114
World War II, 64

X-ray machines, 81

ylem, 61–62, 69, 111

Zwicky, Fritz, 106